Responsive environments

A manual for designers

Ian Bently, Alan Alcock, Paul Murrain, Sue McGlynn, Graham Smith

Responsive environments

A manual for designers

AMSTERDAM • BOSTON • HEIDELBERG • LONDON • NEW YORK • OXFORD
PARIS • SAN DIEGO • SAN FRANCISCO • SINGAPORE • SYDNEY • TOKYO

Architectural Press is an imprint of Elsevier

ELSEVIER

Architectural
Press

Architectural Press is an imprint of Elsevier
The Boulevard, Langford Lane, Kidlington, Oxford, OX5 1GB
30 Corporate Drive, Suite 400, Burlington, MA 01803, USA

First edition 1985
Reprinted 1987, 1992, 1993, 1994, 1995, 1997, 1998 (twice), 1999,
2000, 2001, 2003, 2005, 2006, 2007, 2008 (twice), 2010, 2011

Notice
No responsibility is assumed by the publisher for any injury and/or damage to persons
or property as a matter of products liability, negligence or otherwise, or from any use
or operation of any methods, products, instructions or ideas contained in the material
herein. Because of rapid advances in the medical sciences, in particular, independent
verification of diagnoses and drug dosages should be made

British Library Cataloguing in Publication Data
A catalogue record for this book is available from the British Library

Library of Congress Cataloging-in-Publication Data
A catalog record for this book is available from the Library of Congress

ISBN: 978-0-7506-0566-3

For information on all Architectural Press publications
visit our website at www.elsevierdirect.com

Printed and bound in *China*

11 12 13 14 15   25 24 23 22 21 20

The authors

Alan Alcock is an architect. He has worked in several London offices, including that of Powell and Moya, before forming a practice which developed a special expertise in buildings for health care.

He is now a senior lecturer in the Department of Architecture at Oxford Polytechnic, where he played a central role in introducing the School's uniquely pluralist Diploma course. His current interests include researching the historical development of urban blocks, in their social and economic context.

Sue McGlynn is a town planner and an urban designer. She has worked in a local planning team in the London Borough of Hackney, and has researched into the design and layout of housing estates, with the Social Sciences Buildings Research Team at Oxford Polytechnic.

She is currently a planning officer at Reading Borough Council, mainly working on listed buildings and conservation areas. She has a special interest in urban history, and is currently investigating the historical development of different types of residential layout.

Ian Bentley has practised both as an architect and as an urban designer, in Britain, Holland and the Middle East. He has also spent two years on the board of a property company engaged in both residential and commercial development. He is at present a senior lecturer at the Joint Centre for Urban Design at Oxford Polytechnic, and is a partner in the urban design firm of Bentley Murrain Samuels. With Paul Murrain, Graham Smith and others, he won awards in the RIBA inner city competitions of 1977 and 1980.

His particular interests include designing development strategies for the regeneration of run-down inner-city areas, and researching the effects of the property development process on urban form, building imagery and architectural theory. He has written on these and related topics, in both books and articles, and for television.

Paul Murrain is a landscape architect and urban designer. His experience includes both private practice and work with the Milton Keynes Urban Design Group, of which he was a founder member. He is now a senior lecturer in the Joint Centre for Urban Design at Oxford Polytechnic, and a partner in Bentley Murrain Samuels.

His particular interests include the detailed physical design of outdoor space, and of the interfaces between buildings and the public realm. A musician on keyboards and saxophone, he finds his understanding of musical structure to be valuable in design.

Graham Smith trained as an artist, at St Martin's and at the Royal College of Art, and taught at Goldsmith's College. He has paintings in public and private collections in Britain and France.

He is currently a design tutor in the Department of Architecture at Oxford Polytechnic, with a particular concern for the relationship between architecture and urban design. He has a specialist knowledge of two and three-dimensional geometries; whilst his interest in the architecture of the 1920s and 30s has been expressed through writing, photography and exhibition work.

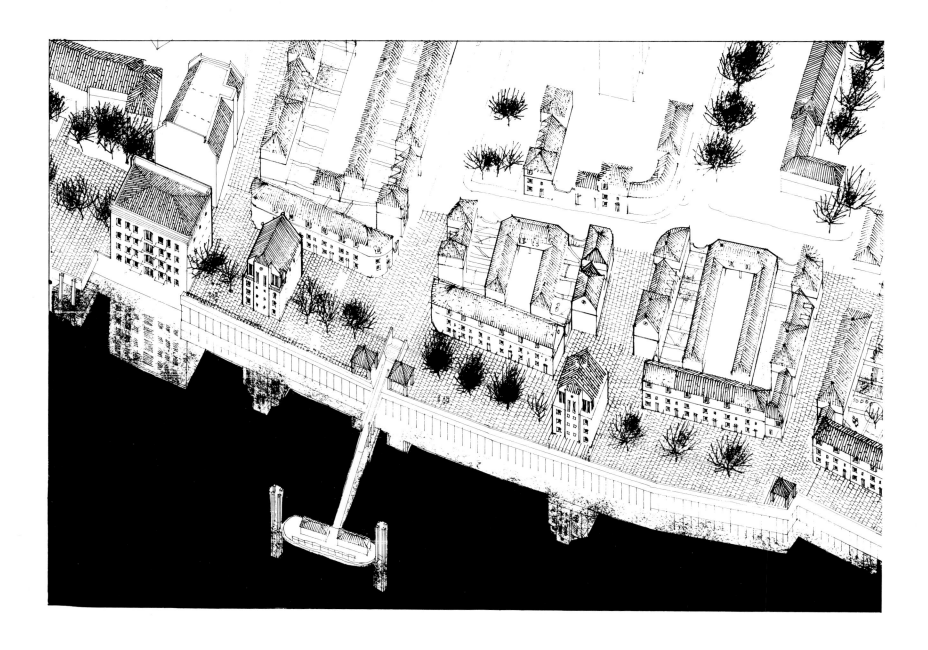

Contents

Acknowledgements

This book grew from our involvement in a series of projects, lectures and seminars under the general title *Responsive Environments*, running in both the Department of Architecture and the Joint Centre for Urban Design at Oxford Polytechnic since 1976. The approach developed in this forum has been tested in practice in various projects, and we owe a debt of gratitude to the planners, architects, engineers, quantity surveyors, developers and estate agents who have coped sympathetically with our sometimes unfamiliar ideas. In this connection, Ian Bentley and Paul Murrain wish particularly to thank their partners, Stef Campbell and Tony Hunt. The project which forms the basis of Chapter 8 was developed in collaboration with Fred Lloyd-Roche, Ken Baker and David Lock, of Conran Roche, architects and planning consultants.

Several of our Oxford Polytechnic colleagues have been especially generous with time and advice: we are particularly grateful to Richard Anderson, to Mike Jenks of the Social Sciences Building Research Team, to Gordon Nelson of the Department of Architecture, to Brian Goodey, Richard Heyward and Ivor Samuels of the Joint Centre for Urban Design, and to Bob Bixby of the Department of Town Planning. Barrie Greenbie of the University of Massachusetts read drafts and made many useful comments.

From the world of estate agency, we are especially grateful to Peter Gibson and David Massif of Gibson Eley and Co., and to Michael le Gray; whilst David Bell of J.T. Developments, and John Foulerton of the North British Industrial Association both gave invaluable advice and encouragement.

Discussions with students have been central to the development of our ideas. Various students have undertaken studies specially to develop and test material for the book itself: we particularly wish to thank Glenn Almack, Robert Ayton, Douglas Brown, Francis Brown, Mike Cheesbrough, Alison Coaker, Basil Constantatos, Brian Curtis, Kjell Dybedal, Nicky Duckworth, Barrie Gannon, Marianne Grand, Ron Morgan, Ian Parry, Laura Rico, Nick Thompson, Andy Trotter and Richard Welch.

The various drafts and the final manuscript have all been typed by Vivienne Ebbs, Stella Thomas and Gillian Long. They often had to work under pressure, and we are very grateful to them. Finally, we must thank the many publishers, copyright-holders and other owners of illustrative material who have generously permitted their use here.

Ian Bentley
Alan Alcock
Paul Murrain
Sue McGlynn
Graham Smith

Introduction

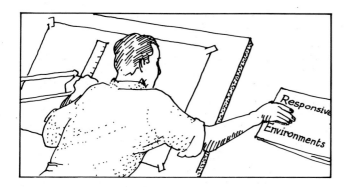

This is a practical book about architecture and urban design. It is meant to be useful on the drawing board, so it does *not* tell you how to do things most designers know already: how to plan buildings efficiently for a given set of activities, how to keep the weather out, how to lay out services and so forth. Designers sometimes do these things rather badly, but at least ways of doing them well are known, and information about them is readily available elsewhere.

We are concerned with those areas of design which most frequently seem to go wrong. As a starting point, we are interested in why modern architecture and urban design are so often criticised as inhuman and repressive, despite the high social and political ideals shared by so many influential designers over the last hundred years.

The tragedy of modern design, it seems to us, is that designers never made a concerted effort to work out the *form* implications of their social and political ideals. Indeed, the very *strength* of their commitment to these ideals seems to have led designers to feel that a concentration on form itself was somehow superficial. Form, they felt, ought to be the *by-product* of progressive social and political attitudes[1]. But in adopting this stance, paradoxically enough, designers failed to realise that the manmade environment is a political system *in its own right:* try walking through a wall, and you'll notice that it is the physical fabric, as well as the way it is managed, that sets constraints on what you can and can't do. Multiplied to the scale of a building or - crucially - a city, this is indeed a political matter.

Once we understand this, it becomes obvious that even from the political point of view it is the things designers do to the built environment that matter. Ideals are not enough: they have to be linked through appropriate design ideas to the fabric of the built environment itself.

This book is a practical attempt to show how this can be done. We start from the same idea as that which has inspired most socially-conscious designers of the last hundred years: the idea that the built environment should provide its users with an essentially *democratic* setting, enriching their opportunities by maximising the degree of *choice* available to them. We call such places *responsive.*

How does design affect choice?

The design of a place affects the choices people can make, at many levels:
- it affects *where people can go,* and where they cannot: the quality we shall call *permeability*.
- it affects the *range of uses* available to people: the quality we shall call *variety*
- it affects how easily people can *understand* what opportunities it offers: the quality we shall call *legibility*.
- it affects the degree to which people can use a given place for *different* purposes: the quality we shall call *robustness*.
- it affects whether the detailed *appearance* of the place makes people *aware* of the choices available: the quality we shall call *visual appropriateness*.
- it affects people's *choice of sensory experiences:* the quality we shall call *richness*.
- it affects the extent to which people can put their *own stamp* on a place: we shall call this *personalisation*.

This list is not exhaustive, but it covers the *key* issues in making places responsive. Our purpose is to show how these qualities can be achieved in the design of buildings and outdoor places.

9

Permeability

Only places which are accessible to people can offer them choice. The quality of *permeability* - the number of alternative ways through an environment - is therefore central to making responsive places.

Permeability has fundamental layout implications. In the diagram below, the upper layout offers a greater choice of routes than the lower one: it is therefore more permeable.

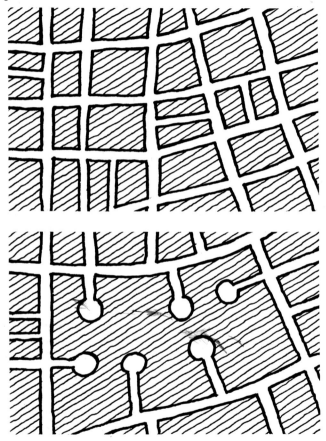

Because it is so basic to achieving responsiveness, permeability must be considered early in design. The designer must decide how many routes there should be, how they should link together, where they should go and - the other side of the coin - how to establish rough boundaries for blocks of developable land within the site as a whole. This design stage is covered in Chapter 1.

Variety

Permeability is of little use by itself. Easily accessible places are irrelevant unless they offer a choice of experiences. *Variety* - particularly variety of *uses* - is therefore a second key quality.

The object of this second stage in design, which is covered in Chapter 2, is to *maximise* the variety of uses in the project. First we assess the levels of demand for different types of uses on the site, and establish how wide a mix of uses it is economically and functionally feasible to have. Then the tentative building volumes already established as *spatially* desirable are tested to see whether they can feasibly house the desired mix of uses, and the design is further developed as necessary.

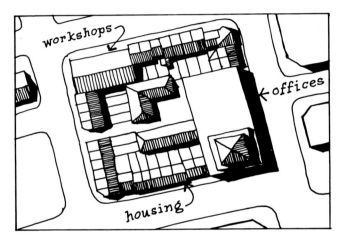

Legibility

In practice, the degree of choice offered by a place depends partly on how *legible* it is: how easily people can understand its layout. This is considered in the third stage of design.

The tentative network of links and uses already established now takes on three-dimensional form, as the elements which give perceptual structure to the place are brought into the process of design. As part of this process, routes and their junctions are differentiated from one another by designing them with differing qualities of spatial enclosure. By this stage, therefore, the designer is involved in making tentative decisions about the *volumes* of the buildings which enclose the public spaces. This process is discussed in Chapter 3.

Robustness

Places which can be used for many different purposes offer their users more choice than places whose design limits them to a single fixed use. Environments which offer this choice have a quality we call *robustness*. This is the subject of Chapter 4.

By this fourth stage in design, we have begun to focus on individual buildings and outdoor places. Our objective is to make their spatial and constructional organisation suitable for the widest possible range of likely activities and future uses, both in the short and the long term.

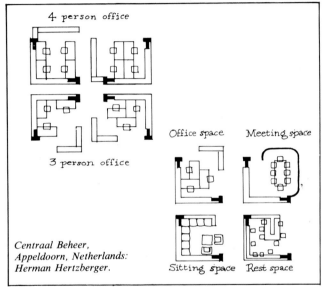

Centraal Beheer, Appeldoorn, Netherlands: Herman Hertzberger.

Visual appropriateness

The decisions we have already made determine the *general* appearance of the scheme. Next we must focus on what it should look like in more detail.

This is important because it strongly affects the *interpretations* people put on places: whether designers want them to or not, people *do* interpret places as having *meanings*. A place has *visual appropriateness* when these meanings help to make people aware of the choices offered by the qualities we have already discussed.

Designing for visual appropriateness forms the subject of Chapter 5. First a vocabulary of visual cues must be found, to communicate the levels of choice already designed into the place. The appearance of the project is then developed in detail, using these cues as the basis for design.

Richness

The decisions about appearance already discussed still leave room for manoeuvre at the most detailed level of design. We must make the remaining decisions in ways which increase the choice of sense-experiences which users can enjoy. This further level of choice is called *richness:* it is the concern of Chapter 6.

By this stage, we are dealing with the smallest details of the project. We must decide whereabouts in the scheme to provide richness, both visual and non-visual, and select appropriate materials and constructional techniques for achieving it.

Personalisation

The stages of design already covered have been directed at achieving the qualities which support the responsiveness of the environment itself, as distinct from the political and economic processes by which it is produced. This is not because we do not value the 'public participation' approach: it is highly desirable. But even with the highest level of public participation, most people will still have to live and work in places designed by others. It is therefore especially important that we make it possible for users to *personalise* places: this is the only way most people can put their own stamp on their environment.

Designing for personalisation is discussed in Chapter 7. Here the designer is making the final detailed decisions about the forms and materials of the scheme; both to *support* personalisation, and to ensure that its results will not erode any *public* role the building may have.

Putting it all together

Taken together, Chapters 1 - 7 outline a step-by-step approach to achieving the qualities we have described:

1. *Permeability:* designing the overall layout of routes and development blocks.
2. *Variety:* locating uses on the site.
3. *Legibility:* designing the massing of the buildings, and the enclosure of public space.
4. *Robustness:* designing the spatial and constructional arrangement of individual buildings and outdoor places.
5. *Visual appropriateness:* designing the external image.
6. *Richness:* developing the design for sensory choice.
7. *Personalisation:* making the design encourage people to put their own mark on the places where they live and work.

In practice, things are more complex than this simple step-by-step structure implies: it is constantly necessary to *modify* the emerging design as you think through the implications of each new step. This process is explored, by means of a case-study, in Chapter 8: *Putting it all together.* Here we show the implications of designing to support all the qualities together, in the context of a large inner-city redevelopment.

How to use the book

Each chapter has three parts:
- an introductory section
- a set of design sheets
- a series of footnotes

Each part contains a different level of information.

The introductory sections

Each introduction discusses how to design for the particular quality concerned. Together, these sections give a comprehensive coverage of responsiveness as a whole. If you are not already familiar with the subject, the best way to start using the book is by reading through all the introductions in sequence.

In our experience, it is necessary to consider *all* the qualities, even when designing quite small schemes. But the *proportion* of design effort expended on each quality tends to vary; depending on the particular site, and on the scale of project concerned. Large, complex schemes demand a greater proportion of time on the qualities covered in the *earlier* chapters, whilst smaller projects are usually more concerned with the *later* ones.

The design sheets

Once designing begins, we need the second level of information: a series of *design sheets,* covering the practical implications of achieving the qualities concerned. The sheets are arranged in the order we have found most useful in our own projects.

The notes

It is impossible for a book like this to cover *all* the eventualities which might arise when designing. So footnotes and suggestions for further reading open up a wider information network, for investigating particular topics in greater depth.

And finally.....

The book as a whole explores an *approach* to designing responsive places. It does not dictate a recipe. So it should be used *creatively.* All its ideas are intended as springboards for design, not as straight-jackets on the designer's imagination.

Chapter 1: Permeability

Introduction

Only places which are accessible to people can offer them choice. The extent to which an environment allows people a choice of access through it, from place to place, is therefore a key measure of its responsiveness. We have called this quality *permeability*.

Permeability: public and private

If everywhere were accessible to everybody, physically or visually, there would be no privacy. But one of our basic sources of choice stems from our ability to live both public and private roles. For this capacity to flourish, both public places *and* private ones are necessary.

Of course, public and private places cannot work independently. They are complementary, and people need access across the *interface between them*. Indeed, this interplay between public and private gives people another major source of richness and choice.

Public and private spaces, and the interfaces between them, each have different implications for permeability.

Permeability and public space.

The permeability of any system of public space depends on the number of *alternative routes* it offers from one point to another. But these alternatives must be *visible*, otherwise only people who already know the area can take advantage of them. So *visual* permeability is also important.

Both physical *and* visual permeability depend on how the network of public space divides the environment into *blocks*: areas of land entirely surrounded by public routes. These can vary radically in size and shape, as illustrated below:

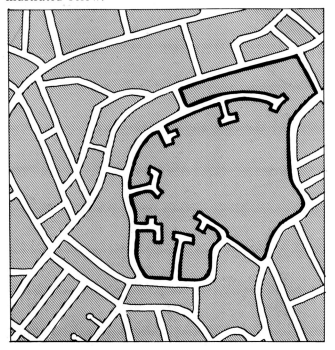

The advantages of small blocks

A place with *small* blocks gives more choice of routes than one with *large* blocks. In the example below, the large-block layout offers only three alternative routes, without backtracking, between A and B. The version with small blocks has nine alternatives, with a slightly *shorter* length of public route.

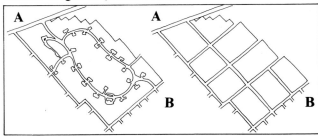

Smaller blocks, therefore, give more *physical* permeability for a given investment in public space. They also increase *visual* permeability, improving people's *awareness* of the choice available: the smaller the block, the easier it is to see from one junction to the next in all directions.

The decline of public permeability

Three current design trends work *against* permeable public space:
- increasing *scale of development*.
- use of *hierarchical layouts*.
- pedestrian/vehicle *segregation*.

Scale of development

Unnecessarily monolithic developments, which could function equally well if divided into smaller elements, produce excessively large blocks.

Hierarchical layouts ✓

Hierarchical layouts reduce permeability: in the example below there is only one way from A to D, and you *have* to go along B and C: never A-D directly, or ADCABCD, but *always* ABCD. Hierarchical layouts generate a world of culs-de-sac, dead ends and little choice of routes.

This is not to say that culs-de-sac are *always* negative: they support responsiveness if they offer a choice which would otherwise be missing. But they must be *added* to a permeable layout, not *substituted* for it.

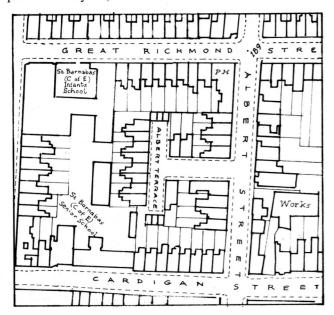

Segregation

Permeability is effectively *reduced* by segregating the users of public space into different categories, such as vehicle users and pedestrians, and confining each to a separate system of routes. When this happens, the only way to give both categories a level of permeability equivalent to a *de*-segregated system is through an expensive duplication of routes.

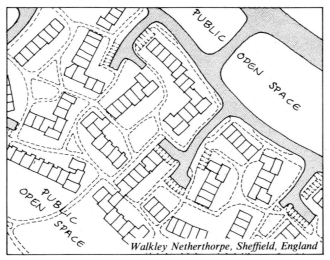

Walkley Netherthorpe, Sheffield, England

Avoid built-in segregation.

Chapter 4 will show other ways of helping motorists and pedestrians to live together. And in any case, it is *never* necessary to build segregation irrevocably into a layout early in design. If we initially make a high level of permeability for *everyone*, then segregation can be achieved later, if necessary, by detailed design or management. This gives future users control over how *they* want to use the place, because they can de-segregate if circumstances change.

Permeability and the public/private interface

Since *physical* access to private space is necessarily limited, permeability across the public/private interface is largely a *visual* concern. This has different implications for public and private space.

The interface: visual permeability

Visual permeability between public and private space can also enrich the public domain. If wrongly used, however, it can confuse the vital distinction between public and private altogether.

This is because not all the activities in private space are *equally* private: there is a gradation, for example, from entrance hall to lavatory. To maintain the public/private distinction, the most private activities must be kept from visual contact with public space.

The interface: physical permeability

Physical permeability between public and private space occurs at entrances to buildings or gardens. This enriches public space by increasing the level of activity around its edges. We shall show how important this is in Chapter 4: for now, it implies that as many entrances as possible should be located round the edges of public spaces, as opposed to what often happens nowadays.

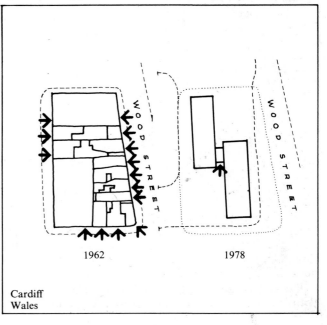

1962   1978

Cardiff
Wales

The need for fronts and backs

This means that all buildings need two faces: a *front* onto public space, for entrances and the most public activities, and a *back* where the most private activities can go. This gives users the chance to do whatever they like in their private space, *including* the right to make rubbish and clutter, without compromising the publicness of public space.

Because private activities out of doors are particularly vulnerable to overlooking, they have to be screened by solid barriers. If they are at the front, adjoining public space, these barriers have a negative, deadening effect, destroying the public character. *Most* of the private outdoor space must therefore be at the back.

Perimeter block development

Applied consistently, the front/back distinction - with private open space at the back, and public open space at the front - leads to a type of layout we call *perimeter block development*.

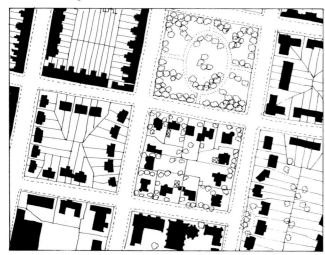

The interface: effects on private spaces

For the public/private interface to make private life *richer,* instead of destroying privacy altogether, it is vital that its degree of permeability is under the control of the private users. Do not worry about this: it is not difficult to achieve at a later stage of design, by using normal building elements like level changes, windows, porches, curtains, sound-reducing glazing and venetian blinds. This will be covered in Chapter 4.

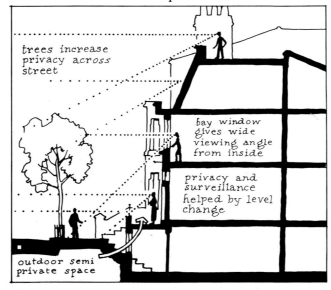

trees increase privacy across street

bay window gives wide viewing angle from inside

privacy and surveillance helped by level change

outdoor semi private space

This degree of control is often not provided nowadays: instead of leaving users to control how much permeability *they* want, designers decide *for* them, by making permanent physical and visual barriers. This is usually because the front/back distinction has been forgotten.

Headington, Oxford, England

Summary: physical form and permeability

The implications of visual and physical permeability make powerful demands on design. The easiest way of meeting these demands is by designing *perimeter blocks:*
- fronts facing outwards onto public space - street, square or park - close enough to enjoy its liveliness.
- backs facing inwards to the centre of the block.
- private outdoor space at the back.

In our experience, other kinds of layouts nearly always lead to permeability problems of one sort or another. It may not *always* be possible to use perimeter block development, but its advantages are so important, and so difficult to achieve in any other way, that it should be considered as the obvious starting point for design.

Starting the design

We have explained the key factors governing permeability, and the reasons why it is a problem nowadays. The next step is to use these ideas in design.

Links to surrounding areas.

In any project large enough to have more than one block, people can potentially move *through* the site from its surroundings, from one side to another.

This choice will only be useful if people are *aware* of it, so it is important to locate new routes as continuations from as many access points as possible outside the site itself, and make sure they can be seen to lead somewhere.

THAT STREET'S NEW... LOOKS LIKE A SHORT-CUT TO THE SHOPS

The first step in design, therefore, is to analyse the layout of routes in the surrounding area; defining the access points onto the site, and noting their relative importance in terms of where they lead to. This is covered in Design Sheet 1.1.

Locating new routes.

This analysis can now be used to position the most important new routes through the site, as discussed in Design Sheet 1.2.

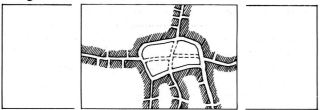

Intensity of use

Now we have located all the routes *through* the site, it is useful to estimate how intensely each is likely to be used by people from *outside* the site. We shall need this information when we come to consider the *uses* in the various blocks, in Chapter 2. For example, high levels of traffic flow might inhibit housing unless handled carefully in detail.

It is easiest to make these estimates while we still have the question of routes in the forefronts of our minds: ways of doing so are covered in Design Sheet 1.3.

Junction design

Next check that the junctions between the proposed streets are acceptable to the traffic engineers. This will depend on the traffic roles of the streets themselves, as discussed in Design Sheet 1.3.

The block structure

The tentative street positions now decided will start to define blocks. These must now be checked for size: make them as small as possible. The minimum practicable size depends on the forms of their perimeter buildings, and on the usage of the private outdoor spaces within the blocks themselves. Both factors are discussed in Design Sheet 1.4.

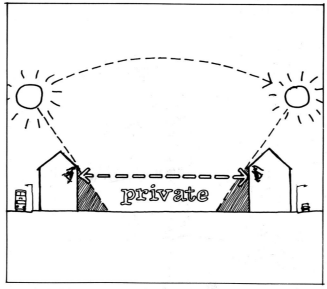

Block shape

Overlooking at the corners can also be a crucial problem, if the blocks are densely built up. This has implications for both planning and massing which can be left till a little later. They are covered in Chapter 3.

Design implications

How to achieve permeability

1. **Analyse the streets and blocks of the surrounding area, to establish the relative importance of all access points to the site (Design Sheet 1.1).**

2. **Locate new routes through the site (Design Sheet 1.2).**

3. **Analyse traffic roles of all the proposed new streets, and check that street widths and junction designs are acceptable to the traffic engineers (Design Sheet 1.3).**

4. **Check that the blocks defined by the new streets are of practicable sizes (Design Sheet 1.4).**

1.1: Using existing links

The starting point for a permeable scheme is the existing system of links into and through the site from the surrounding area. Begin by analysing these links, and deciding how best to use them.

Permeability is important at two scales:
- links which connect the site to the city as a whole
- links which connect the site to its immediate local surroundings

Connections to the city as a whole

To achieve high permeability to and through the site from the city as a whole, we must connect it via the largest possible number of direct links to the system of *main* streets: those carrying through traffic linking the various parts of the city. So begin by finding the nearest main streets beyond the boundaries of your site, marking them on a detailed plan to a scale of not less than 1:10,000.

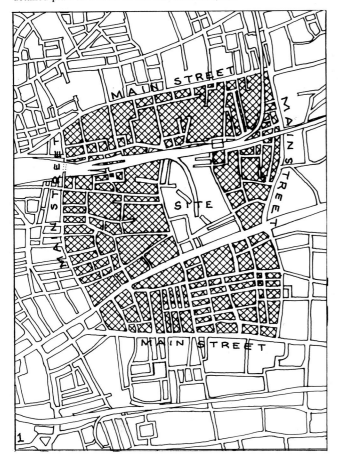

Connections to the main street system

Next, find all the links within this area which connect the site to the system of main streets. Compare them, to see which connect the site *most directly* to the main streets. This can be assessed by comparing the number of changes of viewing point necessary on journeys along each link from the main street system to the site. In the sketch below, link A requires only one change of viewing point and is therefore more direct than link B, which requires three.

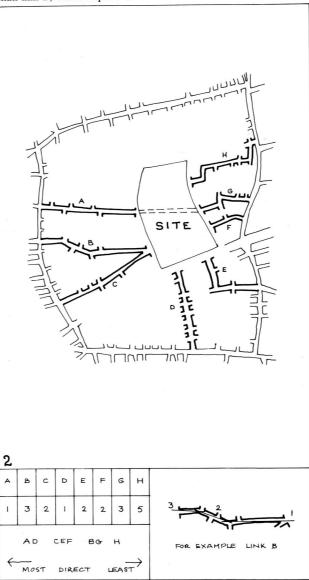

A	B	C	D	E	F	G	H
1	3	2	1	2	2	3	5

AD   CEF   BG   H

← MOST   DIRECT   LEAST →

FOR EXAMPLE LINK B

Connections to immediate local surroundings

Next, within the same area defined by the main streets, consider all the links to the site; including those which do not reach as far as the main streets themselves. Count the number of connections along each one in turn, as shown below. The highest numbers will show which streets link the site most strongly to its immediate surroundings.

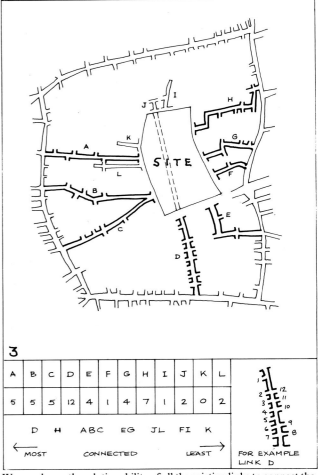

A	B	C	D	E	F	G	H	I	J	K	L
5	5	5	12	4	1	4	7	1	2	0	2

D   H   ABC   EG   JL   FI   K

← MOST   CONNECTED   LEAST →

FOR EXAMPLE LINK D

We now know the relative ability of all the existing links to connect the site both to the city as a whole and to the immediate local surroundings. This information can now be used to decide the relative importance of extending each link into and through the site, to achieve an appropriate balance between permeability at the city-wide and local scales. For instance, in Diagram 2, an east-west route would increase city-wide permeability, whilst in Diagram 3 a north-south street would have more effect on permeability at the local scale.

Once you have decided which links it is most important to extend into and through the scheme, you can begin to align the system of streets and blocks within the site itself, as discussed in Design Sheet 1.2.

1.2: Designing the street/block system

Give users a choice of routes through the site, by keeping perimeter blocks as small as possible.

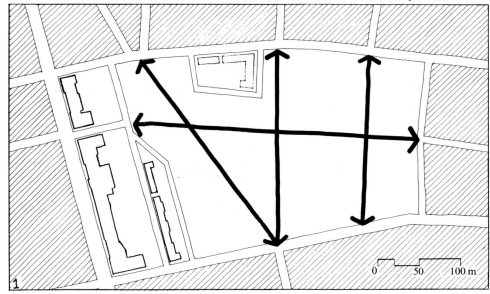

Design sheet 1.1 revealed the most important links to the site. Starting with these, join the access points across the site (1), taking account of any existing routes through it.

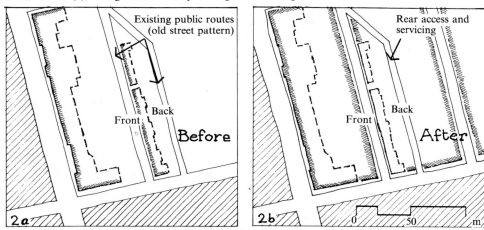

If there are any existing buildings to be kept, note the positions of their fronts and backs. Make sure that public routes run at their *fronts*. (2a,2b).

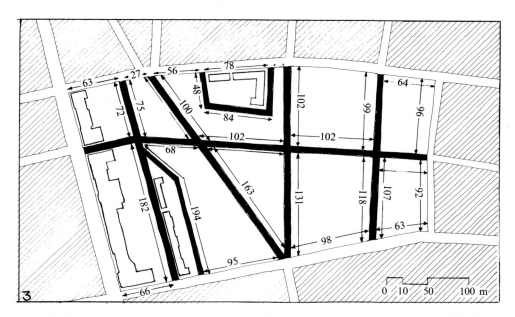

Check the block sizes you have created. Make them as small as practicable, depending on the uses they will house. If you already know these uses, check block sizes with Design Sheet 1.4. If you don't, 80-90 metre blocks will do for most purposes (3). They should only need minor adjustment later, when uses are finally decided.

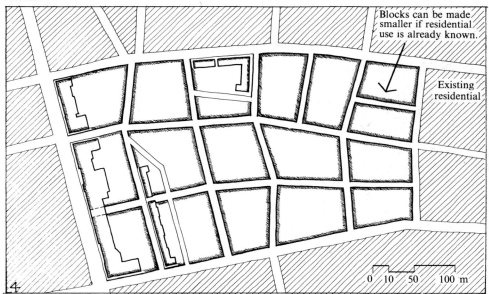

Next, increase the sizes of any blocks which are too small, and subdivide any which are larger than they need be (4) to make the final layout as permeable as possible. Check with Design Sheet 1.3 to see that all the junction designs are feasible.

1.3: Street types and junction design

This Design Sheet shows how to estimate the approximate traffic capacities and carriageway widths of the roads in the scheme, together with their junction designs, prior to detailed consultation with traffic engineers and highway authorities.

Street classification

Both the spacing and the detailed design of junctions depends on the street types they connect. Urban streets are classified by traffic engineers according to their traffic role: the amount and type of vehicular traffic they carry (1). To classify the streets in your scheme, therefore, it is necessary to assess the vehicle flows which each will carry.

1 Urban street types

Design for free-flowing traffic is dominant concern.

All-purpose roads

Primary distributor
Long distance through traffic, serves town as a whole.

District distributor
Through traffic linking main districts within town.

Local distributor
Links traffic within local 'environmental areas'.

Access road
Provides direct access to buildings and land within 'environmental areas'.

Local distributor

Access roads
- Major access
- Collector
- Minor access

Shared surfaces
- Access ways
- Mews courts
- Housing squares

Residential roads

Design for traffic subordinate to environmental factors and Pedestrians' needs.

Estimating vehicle flows

In the case of major roads linking into the main city network, it is necessary either to carry out a traffic survey, or to obtain the relevant flow data from the local highway authority. On streets which carry only local traffic, approximate figures can be calculated from a knowledge of the uses in the buildings and land to which the streets give access.

2 Building use	Vehicles (vph)
Dwellings with 2 or more bedrooms	1 per dwelling
Dwellings with one bedroom	0.75 per dwelling
Elderly persons' dwellings	0.25 per dwelling
Schools:	
pupils up to 12 years	1 per 4 pupils
pupils aged 12 or more	1 per 6 pupils
Places for further education	1 per 2 pupil spaces
Offices	1 per 10sq.m. (gross) or part
Minor or existing warehouse or industrial unit	1 per 5sq.m. (gross) or part
Shopping	1 per 10sq.m. (gross) or part
Commuter car parks	1 per space
Short-stay car parks	2 per space
Churches	1 per 5 seats
Public houses	1 per 2.5sq.m. public area
Clubs, halls and community centres	1 per 5sq.m. (gross)

(Source: Surrey C.C.)

To calculate the approximate flow from table 2, add together the figures for vehicles per hour for all the uses concerned (the exact figures used may vary slightly from one highway authority to another). Figure 3 illustrates a practical application of this technique.

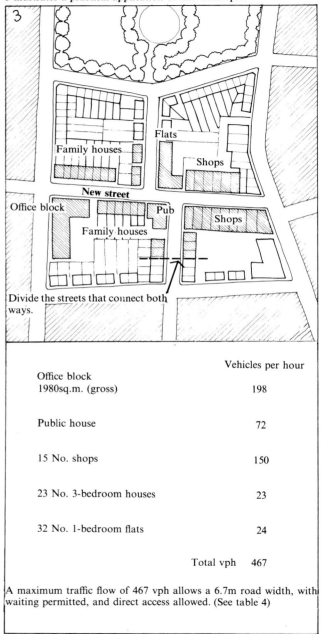

Divide the streets that connect both ways.

	Vehicles per hour
Office block 1980sq.m. (gross)	198
Public house	72
15 No. shops	150
23 No. 3-bedroom houses	23
32 No. 1-bedroom flats	24
Total vph	467

A maximum traffic flow of 467 vph allows a 6.7m road width, with waiting permitted, and direct access allowed. (See table 4)

Carriageway widths

Once the streets have been classified, estimate the carriageway widths required; as shown in table 4.

4a All-purpose roads

Road type	Road width (2 lane, 2 way)		
	6m	6.7m	7.3m
Primary, District and Local distributors, with no frontage access, no waiting, and negligible cross traffic.	1200	1350	1500
	maximum traffic flow vehicles per hour (vph)		
District and Local Distributors and Access Roads with high capacity junctions, but restricted waiting and access	800	1000	1200
Local Distributors and Access Roads with waiting and direct access allowed.	300-500	450-600	600-750

4b Residential roads

Road type	Maximum flow (vph)	width (m)
Local Distributor	400 (min)	6.7
Major access road	300	5.5
Collector road	150	5.5
Access road	45	5.5
Access way	20	4.5
Mews Court	15	4.8
Housing square	15	4.8

Junctions

Next check the spacings between junctions from figure 5. This shows that 90 metre blocks between junctions will be large enough for most situations: we suggest you use this dimension for the block layout even for primary and district distributors. In these cases, traffic will not be permitted to use the intermediate intersections. But the layout will still be permeable for pedestrians; and may eventually be opened up into a more permeable vehicle system, should traffic engineering rules or street roles change in the future.

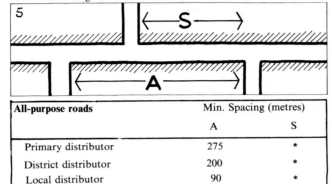

All-purpose roads	Min. Spacing (metres)	
	A	S
Primary distributor	275	*
District distributor	200	*
Local distributor	90	*
Access road	90	*

* No standards exist for this category.

Note: crossroads are allowed in the following circumstances:
- where important roads cross, these generally have to be controlled by traffic signals, roundabouts or grade separation.
- where there are clear priorities, and the minor 'cross' roads are controlled by 'give way' or 'stop' signs.

For junctions with very minor roads, staggered junctions are advised, with spacing S at a minimum of 40 metres.

Residential roads	Min. spacing (metres)	
	A	S
Local distributor	90	40
Major access road	80	40
Collector road	50	25
Access road	30	15
Access way	25	10
Mews court	30	20
Housing square	30	15

In practice, in consultation with traffic engineers, it may be possible to reduce dimension S to zero, at junctions between minor residential roads.

Note that you can only have blocks below 90 metres in streets which only give access to housing: if you have not decided the scheme's uses yet, you will have to re-check the junction spacings after working through Design Sheet 2.1. The classification of street types also affects the detailed design of junctions, as shown in figure 6. Use this information to sketch in maximum building lines at the corners of the blocks.

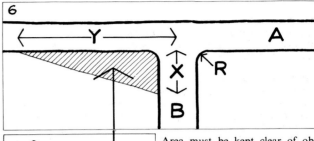

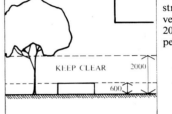

Area must be kept clear of obstructions to visibility, between vertical dimensions of 600mm and 2000mm. (Occasional tree trunks permitted).

All-purpose roads				
Road A	Road B	X(m)	Y(m)	R(m)
Primary distributor	District distributor	9	150 at 50 mph / 120 at 40 mph	
District distributor	Local distributor	9	90	10.5
Local distributor	Access road	9	90	10.5
Access road	Access road	4.5	60	6
Residential roads				
Road A	Road B	X(m)	Y(m)	R(m)
District or local distributor	Local distributor	9	90	10.5
Local distributor	Major access road	4.5	75	10
Major access road	Collector road	4.5	60	9.14
Collector road	Access road	3	60	6
Access road	Access way	2.4	40	6
Access way	Access way / Mews court / Housing square	2.4	30	4.5

1.4: Checking block sizes

The purpose of this Design Sheet is to provide a quick way of checking which uses could be accommodated within the tentative street/block structure already developed. This is an essential preliminary to investigating the demand for different uses, to be covered in Chapter 2.

The minimum size of a perimeter block depends on two main factors:
- the private activities to be housed in the outdoor space within the block: usually private gardens, service access and parking or garaging.
- the form of the buildings around the block perimeter.

Because these factors vary with different building uses, this Design Sheet is divided into three sections, covering the following uses:
- non-residential uses
- flats
- houses with gardens

Each section contains a series of handy reference graphs displaying the relationship between three factors:
- the overall size of the block
- private outdoor space and parking or garaging provision within the block
- characteristics of the buildings around it

The graphs are based on rectangular blocks of the form sketched below. The 'average block dimension' referred to is the mean of two adjacent sides: (A + B) ÷ 2 in the sketch.

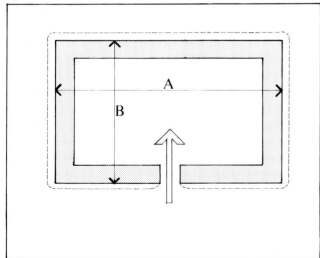

In practice, the dimensions A and B will usually have to be adjusted by a few metres to achieve an efficient parking or garaging layout. At this early stage in design, however, the graphs give an adequate method of checking which uses could be housed in the blocks you are proposing.

Perimeter blocks with non-residential buildings
Worked examples

The graphs are based on continuous perimeter buildings. No allowance is made for space between fronts of buildings and backs of pavements, nor between backs of buildings and parking areas. If you want to include either of these, the average block dimension must be increased as shown below.

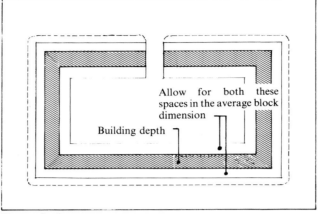

Allow for both these spaces in the average block dimension

Building depth

Example 1 (see pp. 21-23)
Given block size and required parking standard, which building height will enable the maximum area of building to be accommodated in the block?
- Start by locating the relevant block size (1)
- draw a line from it across the graph (2)
- locate the relevant parking standard (3)
- draw a line upwards from it (4)
- the nearest graph line below the intersection of (2) and (4) indicates which building height will achieve the maximum floorspace.

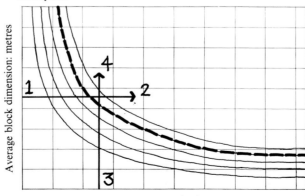

Parking standard:
Cars per gross sq.m. built space

Example 2 (see pp. 21-23)
At a given building height, what is the minimum block size to achieve a given parking standard?
- Start by locating the desired parking standard (1)
- draw a line upwards from it to intersect the relevant building height line at (2)
- draw a line across from (2) to read off the minimum practicable block dimension (3).

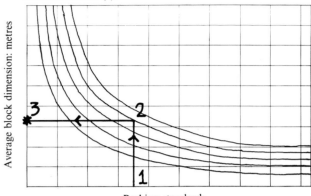

Parking standard:
Cars per gross sq.m. built space

Example 3 (see pp. 21-23)
Given block size and building height, what parking standard can be achieved?
- Start by locating the relevant block size (1)
- draw a line across to intersect the relevant building height line at (2)
- draw a line down from (2) to find the parking standard which can be achieved (3)

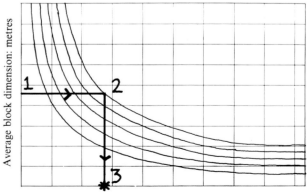

Parking standard:
Cars per gross sq.m. built space

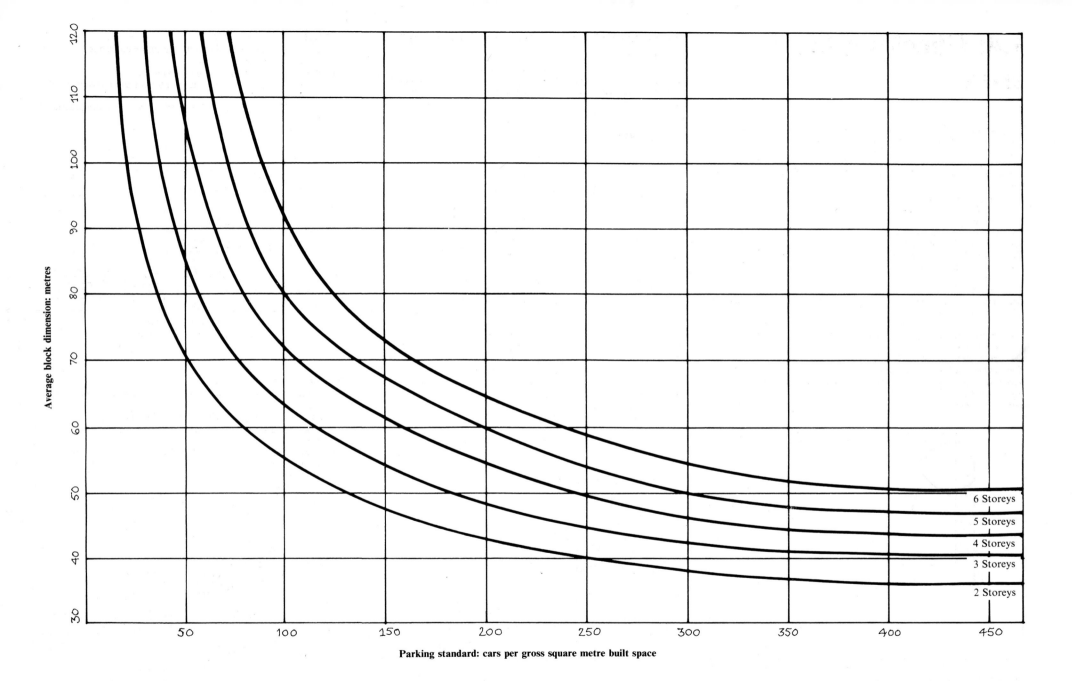

Average block dimension: metres

6 Storeys

5 Storeys

4 Storeys

3 Storeys

2 Storeys

Parking standard: cars per gross square metre built space

Block sizes: non-residential buildings

Perimeter blocks with flats

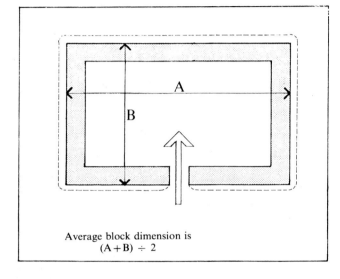

Average block dimension is
(A + B) ÷ 2

Worked examples

The graphs make no allowance for gardens or parking spaces at the front, or for private outdoor spaces within the block. If you want to include these, the average block dimension must be increased as shown below:

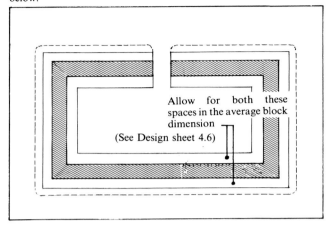

Allow for both these spaces in the average block dimension
(See Design sheet 4.6)

Example 4 (See p23)

Given block size, garden area and required parking standard, which flat size will enable the maximum number of dwellings to be accommodated in the block?

- Start by locating the relevant block size (1)
- draw a line from it across the graph (2)
- locate the relevant parking standard (3)
- draw a line upwards from it (4)
- the nearest graph line below the intersection of (2) and (4) indicates which flat size to use to achieve the maximum number of dwellings.

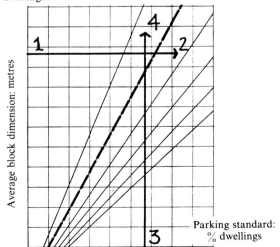

Example 5 (See p23)

Given the area of the desired flat type, what is the minimum block size to achieve a given parking standard?

- Start by locating the desired parking standard (1)
- draw a line upwards from it to intersect the relevant flat area line at (2)
- draw a line across from (2) to read off the minimum practicable block dimension (3)

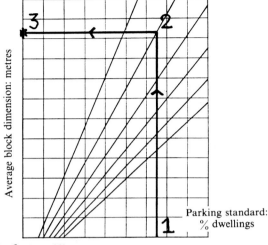

Example 6 (See p23)

Given block size and desired average flat area, what parking standard can be achieved?

- Start by locating the relevant block size (1)
- draw a line across to intersect the relevant flat area line at (2)
- draw a line down from (2) to find the parking standard which can be achieved (3).

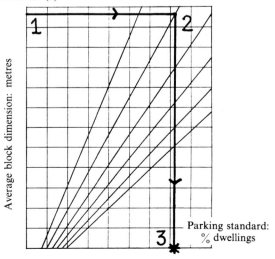

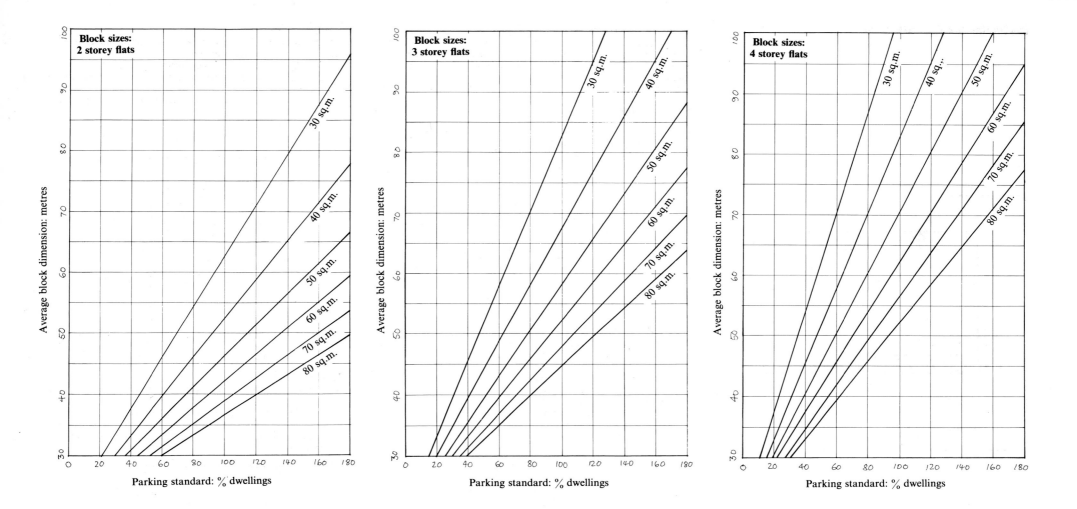

Block size: flats

Perimeter blocks with family houses

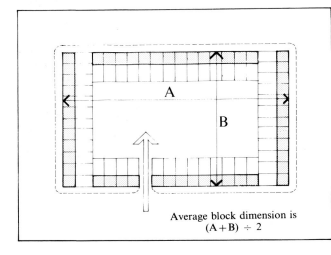

A

B

Average block dimension is
(A + B) ÷ 2

Worked examples

The graphs are based on terraces of 2-storey houses, and make no allowance for front gardens or front parking spaces. If you want to include these, the average block dimension must be increased as shown below.

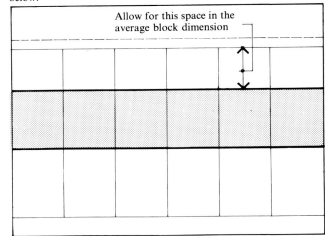

Allow for this space in the average block dimension

Example 7 (See pp. 25-26)

Given block size, garden area and required parking standard, which house type will enable the maximum number of houses to be accommodated in the block?

- Start by locating the relevant block size (1)
- draw a line from it across the graph (2)
- locate the relevant parking standard (3)
- draw a line upwards from it (4)
- the nearest graph line below the intersection of (2) and (4) indicates which house type to use to achieve the maximum number of dwellings.

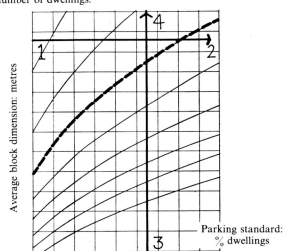

Average block dimension: metres

Parking standard: % dwellings

Example 8 (See pp. 25-26)

Given house type and garden size, what is the minimum block size to achieve a given parking standard?

- Start by locating the desired parking standard (1)
- draw a line upwards from it to intersect the relevant house type / garden size line at (2)
- draw a line across from (2) to read off the minimum practicable block dimension (3).

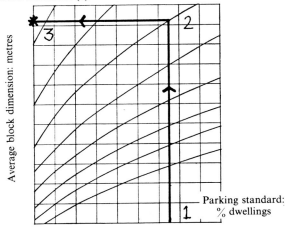

Average block dimension: metres

Parking standard: % dwellings

Example 9 (See pp. 25-26)

Given block size, house type and garden area, what parking standard can be achieved?

- Start by locating the relevant block size (1)
- draw a line across to intersect the relevant house type / garden size line at (2)
- draw a line down from (2) to find the parking standard which can be achieved (3)

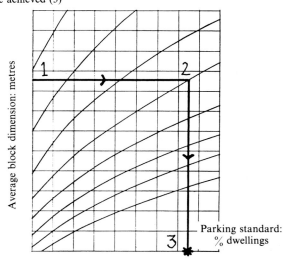

Average block dimension: metres

Parking standard: % dwellings

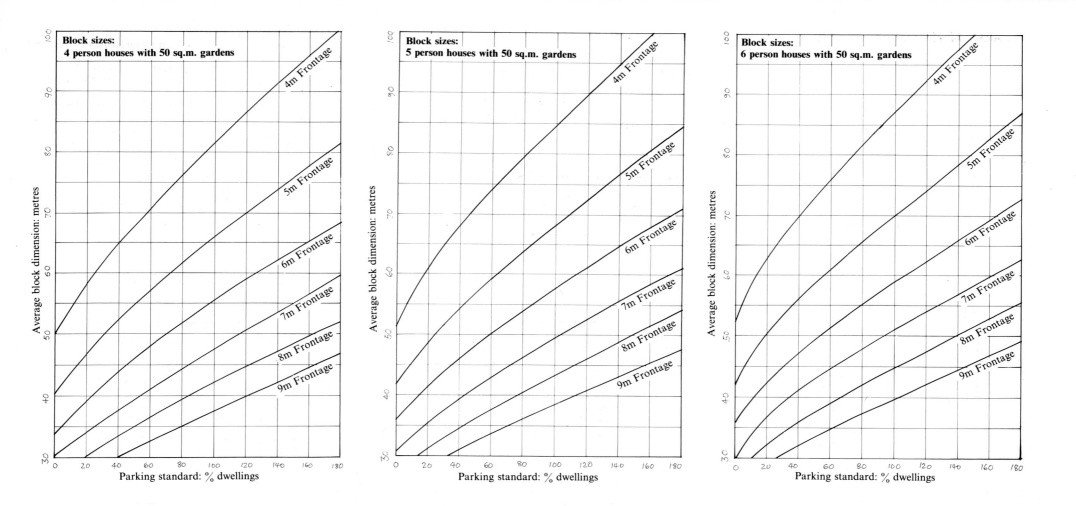

Block sizes: family houses with 50 sq.m. gardens.

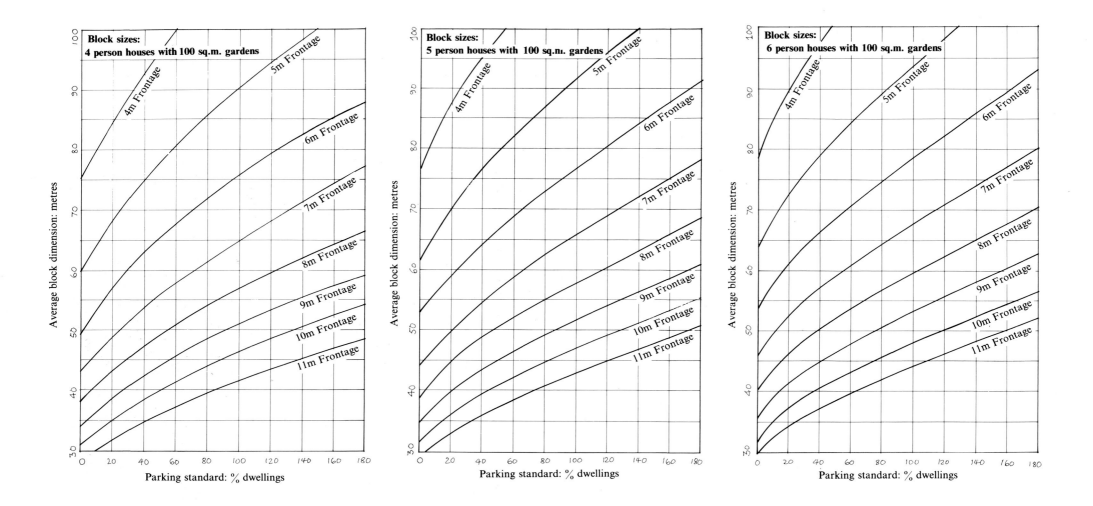

Block sizes: family houses with 100 sq.m. gardens.

Chapter 2: Variety

Purpose

The last chapter discussed how to achieve greater permeability. But accessible places are only valuable if they offer experiential choice: *variety* is therefore the second key quality to be considered.

Different levels of variety

Variety of experience implies places with varied forms, uses and meanings. Variety of *use* unlocks the other levels of variety:

- a place with varied uses has varied building types, of varied forms.
- it attracts varied people, at varied times, for varied reasons.
- because the different activities, forms and people provide a rich perceptual mix, different users interpret the place in different ways: it takes on varied meanings.

Variety of use is therefore the key to variety as a whole. It must be considered early in design,

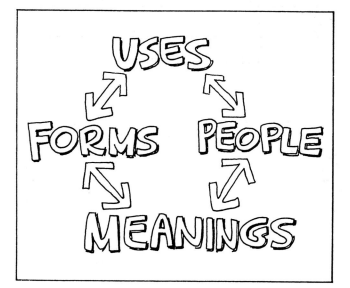

Variety and choice

The purpose of promoting variety is to increase *choice*. But choice also depends on *mobility*: people who are highly mobile can take advantage of a variety of activities even if these are spread over a wide area.

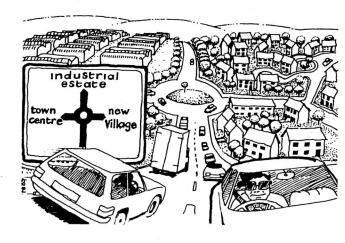

But in practice, who can afford high mobility?[1]
- can children or poor people?
- or disabled or sick people?
- or parents with young children?
- or even women generally?

For people like these - probably the majority, taken together - real choice depends on a close *grain* of variety.

Why is this a problem?

Though their attitudes differ, both developers and planners want *efficient* environments. Developers are interested in economic performance, whilst planners want places which, amongst other things, are easy to manage. Both see their interests as served by two key concepts: *specialisation* and *economies of scale*. Together, these seriously coarsen the grain of variety[2].

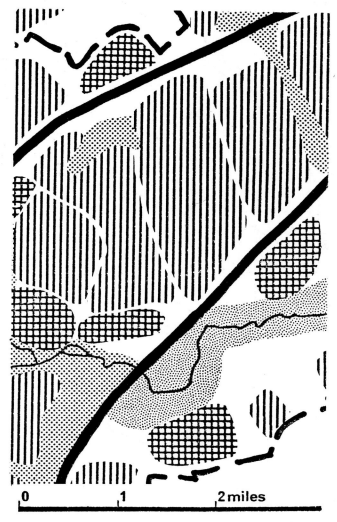

Variety within *districts* is reduced, as they become specialised zones of single use.

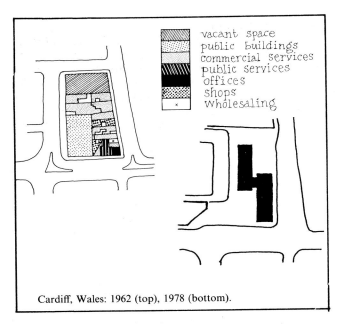

vacant space
public buildings
commercial services
public services
offices
shops
x wholesaling

Cardiff, Wales: 1962 (top), 1978 (bottom).

Variety within *blocks* is reduced, as sites are amalgamated into larger units.

Queen St, Oxford: 1936(top) 1983 (bottom)

Variety within *buildings* is reduced, in the interests of easy management and corporate image.

How much variety?

With all these pressures *against* variety, it is pointless to agonise over exactly how much is needed: designers should simply get the most they can. Because of all the constraints, there is no danger of ending up with too much.

How to maximise it

The variety of uses a project can support depends on three main factors:

- the range of activities which want to locate there, which we shall call *demand*.
- the possibility of supplying *affordable space* in the scheme to house these activities.
- the extent to which the design encourages *positive interactions* between them.

Demand

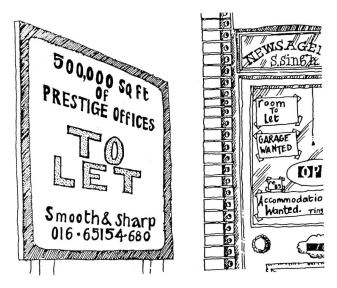

Nobody will fund a project unless it can be shown that there is demand for the space it will provide. Most development agencies, public as well as private, concentrate on a relatively small range of uses for which an *obvious* demand exists.

To draw developers beyond obvious uses, towards greater variety, designers *must* produce a convincing analysis of demand. This design skill, fundamental in promoting variety, is discussed in Design Sheet 2.1

Affordable space

No matter how much an activity wants to locate in a project, it cannot do so unless space is available at a price it can afford.

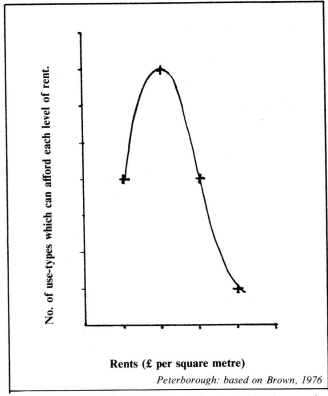

Peterborough: based on Brown, 1976

It is vital to grasp the simple fact illustrated in the graph above: if space in a scheme is available cheaply, many types of users can afford to locate in the place. If space is expensive, only a few types will be found.

To balance the books, schemes which are costly to develop will have to charge relatively high rents or purchase prices. They will be too expensive for most users.

Conversely, to encourage variety we must keep rents and purchase prices low. One important way of doing this is by keeping down the *costs* of the scheme. A wealth of information on cost control exists elsewhere, so we shall not deal with it here[3].

A second way of providing cheap space is by finding a source of *subsidy*.To most designers, this is a less familiar topic.

Subsidies

Subsidies may come either from outside the scheme, or from using some profitable element within it to subsidise other uses which cannot afford an economic rent. This second approach - called *internal cross subsidisation* - is illustrated by design briefs for New York's theatre district. Here, developers are encouraged to build extra offices, provided that a proportion of the extra profits goes to build theatre space, as shown below.

External and internal subsidies are discussed in Design Sheets 2.1 and 2.6 respectively.

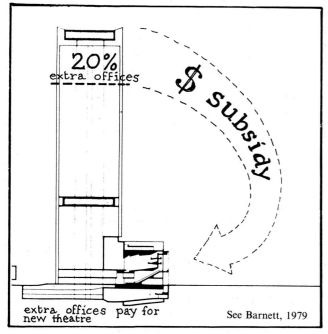

See Barnett, 1979

The role of old buildings

A third way of getting cheap space is by keeping suitable old buildings. When these were built, costs were relatively low, and little money is now tied up in them. Also, there is often limited demand from prosperous tenants to rent old buildings, because these lack modern facilities and offer no boost to tenant prestige. All these factors keep rents low[4].

Because of current construction costs and interest rates, however, even modest new buildings have to charge relatively high rents to break even. Redevelopment therefore implies that rents will rise considerably.

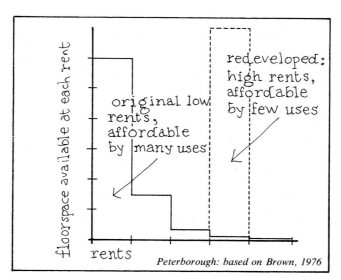

Peterborough: based on Brown, 1976

These increased rents reduce variety, as shown above. This is because the specialised uses which contribute so much to variety are often relatively unprofitable. No matter how well located, users like these cannot afford high rents. Even if given *free* space in a redevelopment, it would probably pay the enterprise to sell its valuable premises, move somewhere cheaper, and pocket the difference . Thus redevelopment, with its high rents, reduces variety. Ironmongers and greengrocers give way to jewellers and offices, as shown below.

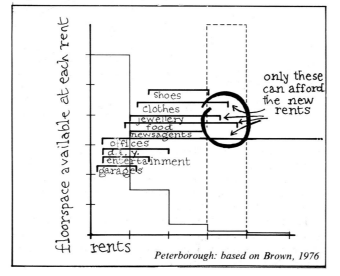

Peterborough: based on Brown, 1976

Design implications

Total redevelopment is bad for variety because all rents are pushed up. But total lack of development is also far from ideal, because it implies that the area will attract neither up-market uses, nor those which need the functional advantages of new buildings. If we want areas of real variety - greengrocers *and* jewellers, for example - then we need a broad spread of rents, as illustrated below.

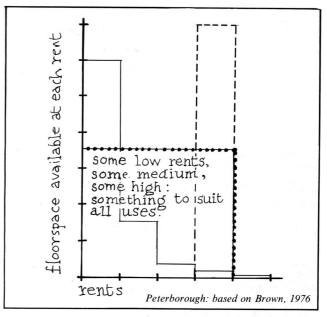

Peterborough: based on Brown, 1976

Thus old buildings should not automatically be kept. They must be carefully selected, to increase variety by housing uses for which demand actually exists. This has two implications:

- their *layout* should be appropriate to the uses concerned[5].
- their *condition* should be suitable for upgrading to an appropriate standard for the uses concerned, at costs which they can afford[6].

A balance of building age and condition - hard to achieve in practice - will generate a variety of rents, supporting a variety of uses. This will be sustainable over time: as the worst buildings are gradually rebuilt, others age and decline in condition; but the presence of a good proportion of medium-to-high rent stock discourages total redevelopment.

Interaction between activities

Variety is not achieved merely by dumping a mixed bag of activities on a site. To work well, the uses should give each other mutual support.

Mutual support

Some activities - *primary uses* - act like magnets, attracting people to a site. Concentrations of dwellings or workplaces are primary uses: nearly everyone *has* to go home, and to work, at frequent intervals. Large stores or markets have a similar effect: many people go to them quite often. In contrast, *secondary uses* are enterprises which themselves lack the pulling-power to attract people, but which live off the people drawn to the place by its primary uses.

Primary uses therefore *support* secondary uses, irrigating them with the pedestrian flows they need for survival. A simple example is the way a shopping centre works. The primary stores attract large numbers of people to the complex, whilst the smaller secondary enterprises - necessary for variety - feed off the pedestrian flow between these main magnets[7].

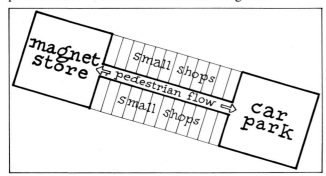

The economic importance of pedestrians explains why shops, for example, pay more for sites with high pedestrian flows. In the diagram below, notice the corner sites; valuable for their *two* streams of pedestrians. (For design implications, see Design Sheet 2.2)

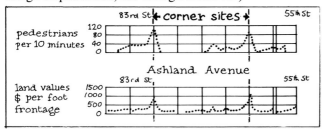

The time element

The time element is also important to this system of mutual support. Some secondary uses - often the most convivial ones, such as pubs and restaurants - need long working hours, perhaps from mid-morning till late evening, to make a living. They are obviously helped if their associated primary uses also draw people into the area over a long period. This usually requires a *mixture* of primary uses. Because of the way in which most people's time is split between work and home, a mix of work space and dwellings functions well in these terms. The modern zoned city fails dismally.

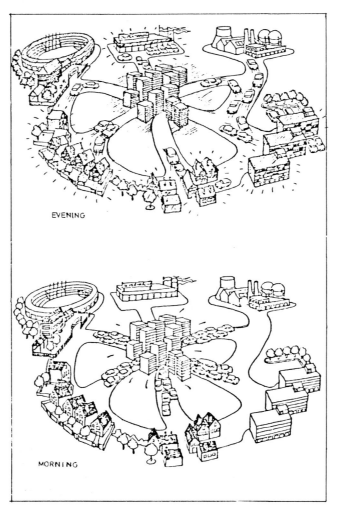

Feasibility

A project's pattern of uses arouses particularly strong interest amongst those with power over the environment, because it is both the basis of economic performance, and a key concern of planning control. In proposing *variety* of use, we are encouraging those with power over the scheme - developers and local authorities - to depart from their usual norms. This will only happen if we can present a convincing demonstration of the project's *feasibility*, at three main levels:
- *functional* feasibility
- *political* feasibility
- *economic* feasibility

Functional feasibility

Some uses are incompatible because of factors like noise or traffic generation. These cannot be located close together. Other uses, however, are incompatible only because people see them as different in status. This problem can often be overcome by careful detailed design.

The potential *advantages* of mixing uses together, when status conflicts can be overcome, are illustrated below: the blank wall on the right, typical of small studio workshops, ensures total privacy for this urban garden.

It is important to pre-empt likely criticisms, from both planners and developers, before they have a chance to harden. This means that we need detailed design studies showing the successful location of any contentious uses proposed, as a key part of the earliest design discussions with either party; as discussed in Design Sheet 2.3.

Political feasibility

Whenever the pattern of uses proposed departs either from accepted norms or from local planning policy, agreement from the local authority will depend at least partly on evidence of public support for the uses put forward. If the initial demand survey (Design Sheet 2.1) has been properly done, the pattern of uses should - at least in part - reflect such local demand as may exist. It is important to demonstrate, as strongly as possible, the support of the local interests which make up this demand.

Economic feasibility

To be economically viable, a project must fulfil one basic condition: its economic value when completed must be greater than or equal to the cost of producing it, plus any profit required by the developers concerned.

To stand any chance of breaking through the low-variety norms current in today's projects, it is *essential* for designers to understand the calculations which are necessary for establishing financial viability. These involve working out costs and values, and amending the design if necessary to balance the one against the other. These topics are covered in Design Sheets 2.4, 2.5 and 2.6 respectively.

But establishing economic feasibility is not merely a matter of juggling figures until they balance on paper. It is crucially important to identify *actual development agencies* willing and able to implement the scheme. There is a limited range of development agencies in existence, each with its preferred types of project and methods of operation. These must be taken into account in the scheme, as discussed in Design Sheet 2.1: in the last analysis, a proposal must appeal to some actual development agency, as well as making economic sense.

Design implications:

How to encourage variety

1. **Take the block structure from Chapter 1 as the starting point for developing variety.**

2. **Considering the widest appropriate range of uses, assess both demand and agencies which could provide accommodation to meet it (Design Sheet 2.1)**

3. **Locate magnets so that pedestrian flow will foster those uses which need it (Design Sheet 2.2)**

4. **Locate remaining uses to minimise negative interactions between them (Design Sheet 2.3) and check tentative block size decisions made in Chapter 1.**

5. **Calculate all costs of scheme (Design Sheet 2.4)**

6. **Calculate project value (Design Sheet 2.5)**

7. **Check economic feasibility, and commitment of development agencies involved. (Design Sheet 2.6)**

2.1: Establishing uses for the site

The first step in designing for variety is to establish which uses exert a demand for space on the site: there is obviously no point in proposing uses for which no demand exists. The purpose of this Design Sheet, therefore, is to show how to find out about demand.

A new role for designers

Traditionally, decisions about uses to include in a project have been seen as part of the patron's role. But this situation is changing; partly because economic recession, on both sides of the Atlantic, is forcing designers to move increasingly into project promotion in order to maintain workloads. In addition, the lack of development pressure in many inner city areas often leads to a situation where the key design question is 'what uses can be found for this piece of land?'. Nowadays, designers of all kinds find themselves grappling with this sort of issue.

Types of demand

Demand can broadly be divided into two categories:
- economic demand
- social demand

Economic demand is exerted from a wide catchment area. It is met by providing space for a certain *kind* of use; without knowing, perhaps until after the project is completed, exactly which enterprise will occupy it.

Social demand is more local and concrete in character: some known group or organisation requires space for some specific and known purpose.

Start with social demand

In our experience, it is best to begin by considering social demand. The enterprises concerned usually have little power, and can easily be left out of account altogether if they are not considered early in the design process.

Local authorities are good first sources of information about social demand:
- they are patrons for a wide range of social projects themselves.
- their estates departments often receive enquiries for space from local organisations of all kinds.

In addition, they can often put you in touch with local neighbourhood and interest groups. These can also be contacted through other sources:
- local libraries
- local newspapers
- community bookshops

Use this combination of sources to assemble a list of accommodation for which a social demand exists; and do not forget to look for demand for outdoor space as well as for buildings. Quantify amounts of space as far as this is possible, and find out the realistic rents or purchase prices which these bodies can afford. Almost certainly, these figures will be low.

Investigating economic demand

For information about economic demand, we must turn to local estate agents: if possible, talk to several and compare their reactions. But estate agents try to minimise the economic risks associated with development projects. They therefore like projects containing uses for which a large and widespread demand exists, appealing to the widest possible number of tenants. This is not a recipe for variety.

Because of this inherent conservatism, it is counter-productive for the designer to discuss the question of demand without some aids to extending the discussion:
- the tentative design proposals from Chapter 1 will help to provoke a more lateral discussion, more concretely related to the specific site.
- So will the ideas about social demand. But most estate agents will regard this whole approach as 'arty': to retain their respect, it is essential that the analysis of social demand has been done in a hard- nosed realistic way.
- to broaden the discussion further, it is useful to produce a checklist of all the uses *you* think are conceivable for the site; again including those out of doors. This can be based on the CI-SFB list of use categories, with the obviously unlikely ones removed[8].

Using these discussion aids, the following questions must be answered:
- which uses are likely to be attracted to the scheme?
- which development agencies are likely to build space for each use?
- could any of the uses be accommodated in existing buildings, if there are any on the site?
- what ancillary supports does each use require? (eg. car parking)
- what are the maximum and minimum areas of space likely to be demanded for each use?
- how much money can the space for each use be sold or let for? (Remember that British estate agents still nearly always use Imperial measurements: check the units you are using, to avoid huge mathematical errors.)
- if the space is to be let, what yield will it realise in investment terms? (Yield is a measure of the annual return on money invested in the project. For example, a 10% yield implies that the project will annually return 10% of the capital invested in it. YP - which is sometimes quoted instead of yield because it makes for easier calculations - is merely the inverse of yield:

$$YP = 1 \div (yield \times 100).$$

- are any of the uses likely to be supported or inhibited by the presence of any of the other uses? (eg. no agency is likely to be enthusiastic about developing family housing next to a scrap yard.)

While you are at it, ask the estate agent's opinion about how much money the site owners might expect to get for it. You will need this information when assessing the financial viability of your proposals.

Planning controls

Once you have established the range of uses for which a *demand* exists, you must investigate any planning controls on the *supply* of space for each use. Now is the time to discuss your proposals with the local planning authority,

A useful pro-forma

The questions to be answered in assessing both the demand for space, and the planning controls on supplying it, add up to a formidable list. The enquiries can most easily be kept under control if you use a pro-forma like that illustrated below. Recording this complex information in a 'ready reference' form will also help you not to forget it - or ignore it - in the later stages of design.

Site	COURAGES BREWERY, READING : SHEET 3 of 8
Interview with:	Michael le Gray Michael le Gray and Co.(shopping developers) 13 March 1984
Use	Shopping
Possible developers	Major national developers, selling on to major financial institutions.
Use existing buildings?	No. (Except for possiblity of slotting a couple of units into the ground floor of the Yield Hall multi-storey car park).
Ancillary supports	See notes below
Demand: max/min areas	Food store 2000-3000m^2, Department store 4000m^2 (max 2 levels). Remainder not more than 35 No.unit shops, each about 6m x 18m.
Rents	Assume about £340/m^2 (Zone A) for unit shops, and £120/m^2 for magnets.
Yields	Assume 6% pending detailed negotiations with institutions.
Negative interactions	See notes below
Planning controls	See notes below

Notes

Ancillary supports:
1. Good magnets to attract shoppers to this area, which is new for shopping. Must have major food store (2000m^2 +) plus a department store (or Heals/Habitat/ Mothercare consortium, or similar).
2. As much car parking as possible (aim for 1000 spaces), within 100m of magnets, on not more than 3 levels. Important to design parking as attractive entrance to scheme. "This must be where people automatically try to park first". (le Gray).
Negative interactions:
Do not put spec.offices over shops: hard to let, for prestige reasons. Allow about 50% storage/admin areas at first floor level. Make sure shopping streets are lined with shops on both sides.
Planning controls:
Local authority likely to go for shopping, but discuss in detail with Officers. Particularly check vehicle access to service areas.

2.2: Concentrating pedestrian flows

Some uses – notably shops – cannot survive without concentrated pedestrian flows. Our permeable street/block structure encourages easy pedestrian access, but to get *concentrated* pedestrian flows we need extra magnets: facilities like large stores, compact markets or large car parks, which attract large numbers of pedestrians. If you find a demand for such uses in Design Sheet 2.1, use them to concentrate pedestrian flows.

The magnets must be located at such a distance from each other, and from any existing pedestrian concentrations, to make the flow between them available to other uses which need it (1,2). Some idea of the maximum spacing can be gained from shopping centre design: the effective range of magnets is 90 - 120 metres[9].

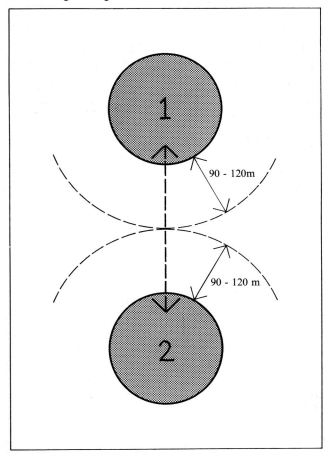

The linking streets must be carefully designed to get the maximum benefit from the pedestrian flow. Make them as narrow as you can (see Design Sheet 1.3) and design them so users can see goods displayed on both sides (3); making sure vehicle traffic is light, so people can easily cross from one side to the other (4).

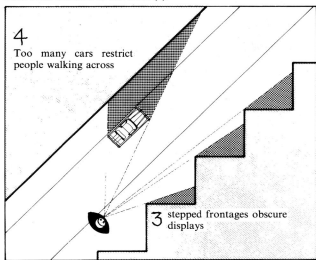

4
Too many cars restrict people walking across

3 stepped frontages obscure displays

Uses needing pedestrian flow must be located on the main inter-magnet links (5). The side streets off these links also have above-average pedestrian flows, but not enough to keep shops alive in high-rent new buildings. But put *robust* buildings here: as buildings grow older, and rents fall, these will attract a wider range of pedestrian-orientated uses (6).

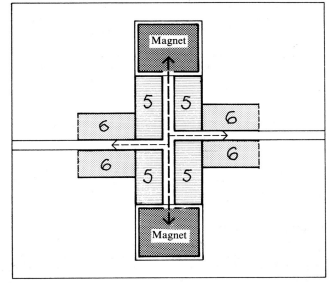

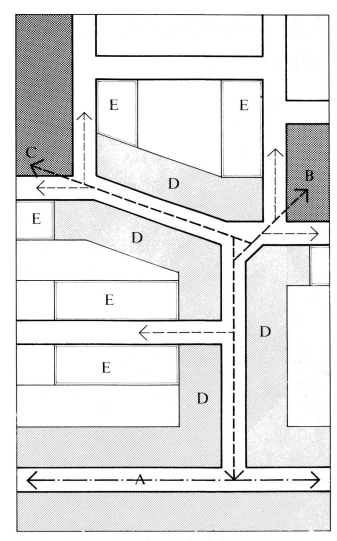

This plan illustrates the process of concentrating pedestrian flows:

1. Locate existing pedestrian concentrations (A)
2. Position the magnet as far from A as practicable (B)
3. Position the second magnet, if there is one, as far from A and B as practicable. (C)
4. Make narrow linking streets between A, B and C, avoiding stepped frontages. Make sure these streets are not likely to carry heavy traffic.
5. Position shops, and other uses needing heavy pedestrian flows, along these streets (D).
6. Make a note to pay special attention to robustness of buildings in streets which link into D. (E)

2.3: Relating incompatible uses

Some uses are incompatible because of functional factors like noise or traffic generation. These cannot be located close together. But other uses are incompatible only because people see them as different in status. This Design Sheet suggests how to relate uses of different status in a small area with the minimum of conflict.

To minimise status conflicts, we must make sure that where different uses directly adjoin one another, they are approximately equal in status. The first step, therefore, is to assess the relative status of the various elements in the scheme. In large projects, this is a complex matter: status obviously varies *between* uses (for example, offices usually have higher status, in the eyes of most people, than do workshops). But often it also varies from front to back of the *same* use (for example, fronts of houses have higher status than their backs).

The easiest way to think this through is by using a matrix like the one below. Remember to include all relevant *existing* uses, either on your own site, or round the edges of your scheme on adjoining sites.

Once you are clear about the compatibilities and incompatibilities between them, the various uses in the scheme can be located as follows:

1 Take a plan, showing the block structure already designed, and note on it the positions of all the existing uses on and adjoining the site.
2 Work round the site, noting in each block all the uses which are compatible with the existing uses adjoining it.
3 By inspection, identify the new use(s) which have least choice of location, and fix them in position.
4 Note the effects of this decision on possible uses for the adjoining blocks.
5 Repeat steps 3 and 4 above, until all the uses have been positioned.
6 Check that this layout enables each use to be reached via streets which themselves contain uses compatible with it, and amend the layout if necessary.
7 If at any stage in this process compatibility cannot be maintained, then the use causing the problem must either be omitted from the scheme, or screened from other uses around it. To avoid disrupting the public street front where this occurs, consider whether the use concerned could be screened by another single-aspect use, to continue the public front. (This topic is discussed in more detail in Design Sheet 4.3).
8 Any incompatibilities which may unavoidably remain can be tackled through the detailed design of the buildings concerned. Chapter 5 will cover this in detail.

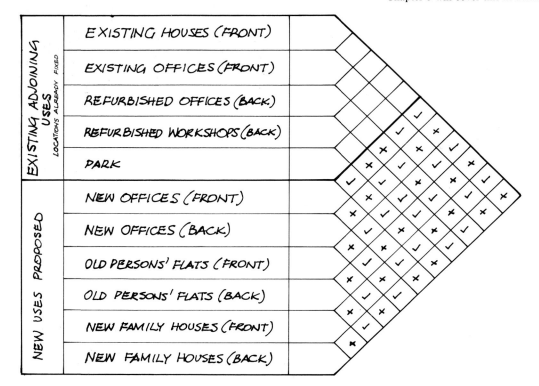

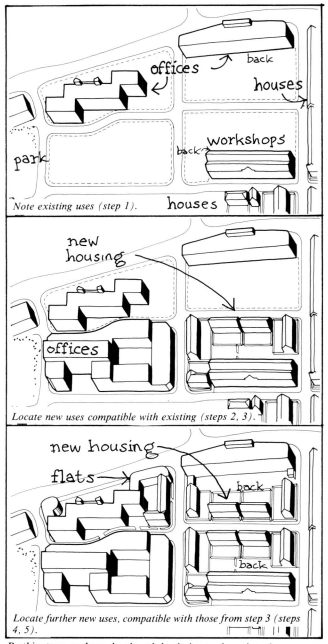

Note existing uses (step 1).

Locate new uses compatible with existing (steps 2, 3).

Locate further new uses, compatible with those from step 3 (steps 4, 5).

By this stage, we have developed the design to the point where we are able to measure the proposed areas of each use. With this information, we can proceed to consider the project's financial implications, as discussed in Design Sheets 2.4, 2.5 and 2.6.

2.4: Calculating project values

The first step in assessing the economic feasibility of a project is to calculate its economic value. This Design Sheet is about ways of working this out; both for elements of the project which are to be sold, and for those which are to be let.

An example

The necessary calculations are best explained with the help of a worked example. The project we shall use to illustrate all the points made in this Design Sheet is sketched below. We shall continue to use the same example in Design Sheets 2.5 and 2.6.

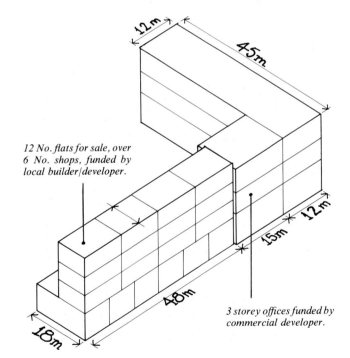

12 No. flats for sale, over 6 No. shops, funded by local builder/developer.

3 storey offices funded by commercial developer.

Project elements to be sold

For project elements to be sold - in this case the flats - the value is equal to the expected sale price. This is easily estimated:

- Step 1:
  establish the numbers of each type of unit to be sold. (12 flats in our example)

- Step 2:
  refer to the estimated selling price for each type, from Design Sheet 2.1. (£28,000 for each flat in our example)

- Step 3:
  simple multiplication will indicate the value:

  Value of flats = 12 × £28,000
  = £336,000

Project elements to be let

There are various ways of calculating the value of those elements of the project which are to be let[10].The approach outlined below is the one most developers use: it is important to use this method because only by doing the *same* calculations as the developer can we predict whether our project will attract the development agency support necessary to make it actually happen. Using this approach, the value of each of the project elements to be let - shops and offices in our example - is calculated from the following formula[11].

£ Value = £ Annual rent roll × 100 ÷ percentage yield

Alternatively if YP rather than Yield is being used:

£ Value = £ Annual rental × YP

The calculation is made as follows:

Step 1

Establish the lettable area proposed. This has different implications for different uses :
- for flats, offices and flatted workshops, the figure to use is 'nett lettable area': the total internal area of the buildings, minus the area used for communal services and circulation. Usually, nett lettable area is approximately 0.8 x total building area.
- for single storey industrial units, or others where there are no corridors to be considered, take nett lettable area as equal to internal building area.
- for shops, the nett lettable area is equal to the internal building area; but it is necessary to divide this into 'zones', called A , B , C , and so on, each 6 metres deep, running parallel to the shopping frontage, as shown in the plan below.

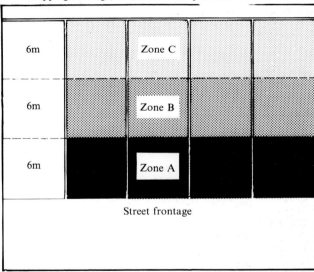

For our example, the lettable areas are calculated as follows:
Offices:
gross area of offices (per floor)
= 12m × (45 + 15)m = 12m × 60m
= 720 sq.m.
therefore total gross area of offices
= 3 (storeys) × 720 sq.m.
= 2160 sq.m.
therefore nett lettable area
= 2160 sq. m. × 0.8
= 1728 sq.m.

Shops:
shop space must be divided into the zones mentioned above. Then;
lettable area (zone A) = 6 × 48 sq.m. = 288 sq.m.
lettable area (zone B) = 6 × 48 sq.m. = 288 sq.m.
lettable area (zone C) = 6 × 48 sq.m. = 288 sq.m.

Step 2:

Look up the estimated rental per unit area for each use, as established in Design Sheet 2.1. This is taken as a flat rate for all the lettable area, except in the case of shops. For shops, start with the annual rental for zone A. Zone B rents are half those for zone A, zone C half zone B, and so on. But note that the yield does not vary like this: it stays the same over the entire nett lettable shop area.

In our example, the annual rentals per sq.m. - which we established in Design Sheet 2.1 - are as follows:

Offices: £100 / sq.m.
Shops (zone A) £200 / sq.m.
Shops (zone B) £100 / sq.m.
Shops (zone C) £50 / sq.m.

Step 3:

Multiply annual rental per unit area × nett lettable area to get the annual rent roll for each use. In our example, the calculations are as follows:

Annual rent roll for offices
= nett lettable area × rent / sq.m.
= 1728 sq.m. × £100 / sq.m.
= £172,800

Annual rent roll for shops:

Zone A: nett lettable area × rent / sq.m.
= 288 sq.m. × £200 / sq.m.
= £57,600

Zone B: nett lettable area × rent / sq.m.
= 288 sq.m. × £100 / sq.m.
= £28,800

Zone C: nett lettable area × rent / sq.m.
= 288 sq.m. × £50 / sq.m.
= £14,400

Therefore, total annual rent roll for shops
= £57,600 + £28,800 + £14,400
= £100,800

Step 4:

Look up the estimated yield (or YP) for each use, as established in Design Sheet 2.1. In our example, the figures are as follows:

Estimated yield for offices: 10 % (or YP = 10)
Estimated yield for shops: 8 % (or YP = 12.5)

Step 5:

Calculate the value of each use, using one of the formulae below:

- Value = annual rent roll × 100 ÷ percentage yield
  or
- Value = annual rent roll × YP

In our example, both these methods are illustrated in the calculations below:

Alternative 1:

Value = annual rent roll × 100 ÷ percentage yield

Value of offices = £172,800 × 100 ÷ 10
= £1,728,000

Value of shops

Zone A: £57,600 × 100 ÷ 8
= £720,000
Zone B: £28, 800× 100 ÷ 8
= £360,000
Zone C: £14,400 × 100 ÷ 8
= £180,000
Therefore, total value of shops
= £720,000 + £360,000 + £180,000
= £1,260,000

Alternative 2:

Value = annual rent roll × YP

Value of offices = £172,800 × 10
= £1,728,000

Value of shops

Zone A: £57,600 × 12.5
= £720,000
Zone B: £28,800 × 12.5
= £360,000
Zone C: £14,400 × 12.5
= £180,000
Therefore, total value of shops
= £720,000 + £360,000 + £180,000 = £1,260,000

Step 6:

Add together the values of the separate uses, to arrive at the value of the project as a whole. In our example, the calculation is as follows:

Value of flats = £336,000
Value of offices = £1,728,000
Value of shops = £1,260,000
Therefore, total value of the scheme
= £336,000 + £1,728,000 + £1,260,000
= £3,324,000

The value of the project is only one of the factors determining whether or not it is financially feasible. Next we must calculate the project's *costs:* this is covered in Design Sheet 2.5.

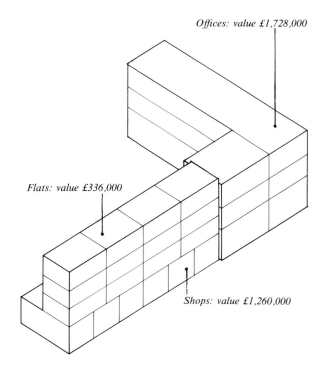

Offices: value £1,728,000

Flats: value £336,000

Shops: value £1,260,000

2.5: Calculating project costs

This Design Sheet covers the second step in assessing economic feasibility: estimating the financial costs associated with the project.

The total cost of getting the project on the ground includes the following main elements:
- site acquisition costs
- construction costs
- professional fees
- cost of short-term funding

We shall illustrate the calculations involved through the same example used in Design Sheet 2.4:

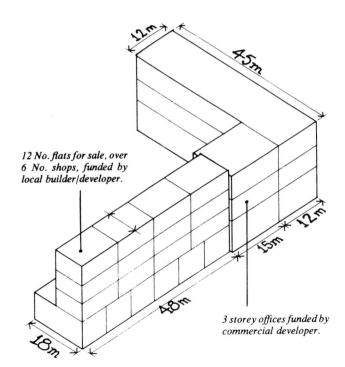

12 No. flats for sale, over 6 No. shops, funded by local builder/developer.

3 storey offices funded by commercial developer.

Site acquisition costs

Establish what the owners of the site expect to sell it for. This is often quite difficult to find out, but one or more of the following approaches should succeed:

- Ask the owners, or their agents, directly: the obvious approach if the site is formally on the market.
- Ask other agents. They can often give a guide based on other land nearby which has recently changed hands. But remember that most estate agents still use Imperial measures, quoting land costs in £ per acre. Make sure you know which units are being used, or you will get bizarre results.
- If any scheme has already been put forward for the site, with the backing of the landowners, you can work backwards to find out the land value they must have assumed. If you find yourself in this situation, skip forward at this point, to the section entitled 'Maximum acquisition costs' in Design Sheet 2.6, where this process is explained with an example.

For our present purposes, we shall assume that our estate agent has advised us that the land will cost approximately £625,000.

Construction costs

Assessing construction costs is the speciality of the Quantity Surveyor, whose advice should be sought if available. If not, the following approach will produce a sufficiently accurate budget figure for these preliminary calculations.

Costs of new construction

First list the gross area of each proposed use, and measure the area of any new street system or other major external works in the scheme. In our example, the gross building areas are as follows:

- Flats:
  3(storeys) × 8m. (building depth) × 48m. (building length)
  = 1152 sq.m.
- Offices:
  3(storeys) × 12m. (building depth) × 60m (building length)
  = 2160 sq.m.
- Shops:
  18m. (building depth) × 48m. (building length)
  = 864 sq.m.
  (note: no new street system or major external works proposed).

Once gross areas have been calculated, look up building costs per unit area for each use, and for streets and external works, using a current builder's price book[12]. Remember that the figures quoted will be for average schemes: take QS advice on amending them if, for example, your site has abnormal ground conditions, or infrastructure problems. Also make any necessary allowances for inflation since the date of publication of the cost figures concerned.

In our example, the figures are as follows:

- Flats: £300/sq.m.
- Offices: £490/sq.m.
- Shops (taken as shells to be fitted out at tenant's cost): £240/sq.m.

Finally, to calculate construction cost, use the following formula:

Construction cost = gross area × cost per unit area.

In our example, the calculations are as follows:

- Flats:

  Construction cost = 1152 sq.m. (gross area) × £300/sq.m.
  = £345,600

- Offices:

  construction cost = 2160 sq.m. (gross area) × £490/sq.m.
  = £1,058,400

- Shops:

  construction cost = 864 sq.m. (gross area) × £240/sq.m.
  = £207,360

Therefore total construction cost = £1,611,360.

Cost of altering existing buildings

The cost of alteration work is far harder to estimate, because there is so much variation in layout and condition between one existing building and another. First carry out a condition survey, and then use QS advice or appropriate specialist references.

Cost of professional fees

This covers the cost of employing architects, quantity surveyors, structural and services engineers, estate agents and legal advisors. Most of these calculate their fees as a percentage of construction costs.

To get an accurate fee estimate, first list the likely advisors needed, then look up the fee scales of the relevant professional institutions, and apply the appropriate percentages to the construction cost you already worked out. As a rough rule of thumb for new work, a total fee expenditure of 12 - 15% of the construction cost will be near enough for these early calculations.

In our example, we shall assume fees at 15% of construction costs:

Cost of fees = £1,611,360 (construction cost) × 15/100 = £241,704

Cost of short-term funding

This is the cost of borrowing money to acquire the land, and get the project built, during the period before it can start producing any income. It is calculated on the average borrowing during the borrowing period. This is difficult to estimate accurately, but a guide is given by the graph sketched below, which shows (very roughly) how the short-term borrowing varies during the construction period.

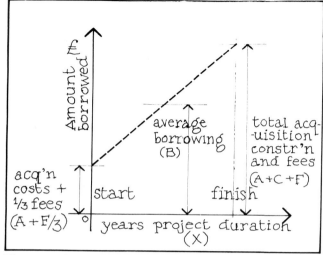

The average borrowing (B) is as follows:

B = (Initial borrowing + Final borrowing) ÷ 2.
By inspection, Initial borrowing (IB) = A + F/3
and Final borrowing (FB) = A + C + F
So B = (IB + FB) ÷ 2
= (A + F/3 + A + C + F) ÷ 2
= A + C/2 + 2F/3

Then cost of short-term funding = B × years construction period × compound interest rate payable.

The interest rate will vary from time to time. It will be set at some amount higher than the bank's Base Rate, by an amount which varies according to the type of project and the status of the borrower. Your bank manager will be able to suggest a realistic figure for your particular project.

In our example, let us assume an interest rate of 12% compound, and a construction period of 2 years.

As we have seen, Average Borrowing = A + C/2 + 2F/3
= £625,000 + (£1,611,360 ÷ 2) + (2 × £241,704 ÷ 3)
= £625,000 + £805,680 + £161,136
= £1,591,816
As we have seen, cost of short-term funding
= Average Borrowing × years construction period × % compound interest rate ÷ 100.
= £1,591,816 × 2 × 1.2544[4]
= £399,355

Total cost

Having calculated the costs of acquisition, construction, fees and short-term funding, add them all together to arrive at the total cost of getting the project on the ground[14]:

Cost of acquisition	= £625,000
Cost of construction	= £1,611,360
Cost of fees	= £241,704
Cost of short-term funding	= £399,355
Therefore total project cost	= £2,877,419

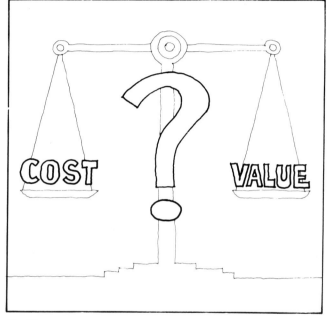

Profit

If the patron for your project is a private-sector development agency - as in our example - then you will have to make an allowance for the developer's profit. This is taken as a percentage of the total project cost calculated above. The actual percentage will vary according to the state of the development market, but is usually somewhere between 15% and 20%: if the profit is lower than 15%, then the chances are that you will have difficulty making the project happen. At 20%, you can be fairly sure of success.

For our example, let us aim for a developer's profit of 20%:
Profit required = total project cost × profit% ÷ 100
= £2,877,419 × 20 ÷ 100
= £575,484

By this stage, we have thought the project through to the point where we have estimated, in financial terms, both its value and its cost. Now we can relate these two figures together, to assess whether or not the project is financially viable. We shall show how to do this in Design Sheet 2.6.

2.6: Checking economic feasibility

The project will be financially feasible if its finished *value* (plus any extra subsidies that can be attracted to the scheme from outside) is greater than or equal to all the *costs* involved in producing it (plus any necessary developer's profit). In Design Sheet 2.4 we calculated the value of the project, and in Design Sheet 2.1 we noted whether any external subsidies were likely to be available. So now we have all the information we need to check the project's financial viability.

Example

A project is only financially feasible if
(Value + subsidies) > or = (costs + profit)

In our example in Design Sheets 2.4 and 2.5, we calculated figures for value and costs as follows:

value = £3,324,000
subsidies = £0
therefore (value + subsidies) = £3,324,000

costs = £2,877,419
profit = £575,484
therefore (costs + profit) = £3,452,903

Since (value + subsidies) is less than (costs + profit), the project is not financially feasible as it stands. But it only falls short by £128,903: what, if anything, could we do to retrieve the situation?

How can we make the project feasible?
There are three main ways of making the project more viable:

- altering the mix of uses to include more profitable elements.
- putting more accommodation on the site.
- rethinking the factors which affect value and cost: rents, yields, acquisition cost, building cost, fees, short-term funding and profit.

Altering the mix of uses
Some of the uses in our example are more profitable than others. We can see this by calculating the profit/sq.m. of each use, as in the table below:

	Offices	Shops Zone A	Shops Zone B	Shops Zone C	Flats
construction cost (£/sq.m.)					
	490	240	240	240	300
fees at 15% construction cost					
	73.5	36.0	36.0	36.0	45.0
short-term funding (£/sq.m.) (from design Sheet 2.5)					
	70.6	34.6	34.6	34.6	43.2
total cost (£/sq.m.)					
	634.1	310.6	310.6	310.6	388.2
value (£/sq.m. nett) (from Design Sheet 2.4)					
	1000	2500	1250	625	573.4
profit (£/sq.m. nett)					
	365.9	2189.4	939.4	314.4	185.2

The bottom line of the table shows the profit contributed to the project by each square metre of each use. The viability of the scheme can be improved by replacing less profitable uses by more profitable ones.

For example, the profitability of flats is £185.2/sq.m., whilst for offices it is £365.9/sq.m. So every sq.m. of flats we replace by offices will generate an extra profit of £365.9 - £185.2 = £180.7

Now, the extra profit required to make the scheme viable is £128,903. This can be achieved by replacing 128,903 ÷ 180.7 = 713 sq.m. of flats by offices, to produce a scheme like that sketched below.

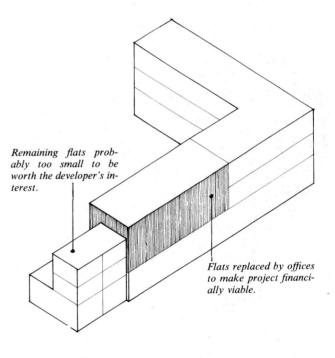

Remaining flats probably too small to be worth the developer's interest.

Flats replaced by offices to make project financially viable.

In practice, this would reduce the area of flats to such an extent that probably no development agency would be interested in them: variety would be reduced. If this is the case, we shall only be able to keep the relatively unprofitable elements of the scheme by adopting a different approach: generating more profit by increasing the *total* area of building on the site.

Putting more on the site

We can calculate the additional area of offices necessary to keep the flats, as follows:

- Extra profit per extra sq.m. of offices = £365.9
- Extra profit needed to make scheme viable = £128,903
- Therefore extra area of offices needed
  = extra profit needed ÷ extra profit per sq.m. offices
  = 128,903 ÷ 365.9
  = 352 sq. m.

This would imply changing the form of the scheme as sketched below.

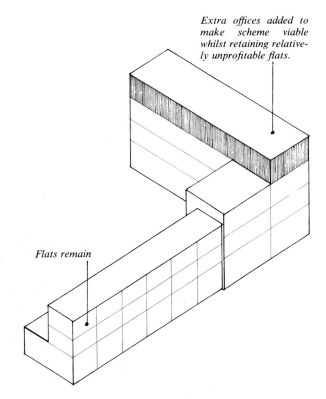

Extra offices added to make scheme viable whilst retaining relatively unprofitable flats.

Flats remain

Provided that planning permission would be granted for these extra offices, and given that they could successfully be let, this approach has created a financially viable scheme which still maintains variety by keeping the entire mix of uses for which a demand was established. Clearly the same approach can be extended to create so-called 'planning gain': subsidising unprofitable uses by building a greater area of the profitable ones.

For example, suppose that we have found a demand for a local community hall, and established that money could be raised for running costs, but not to pay any rent: effectively, therefore, the hall's commercial value is nil. Despite this, the hall might still be funded; by further increasing the area of profitable uses on the site. For example, we can calculate the extra area of offices which would be needed to fund the hall, as follows.

Let us assume that the extra cost of the scheme due to the hall (including construction costs, fees and short-term funding) is £150,000. In addition, let us assume that the hall takes up space previously occupied by one shop and one flat; so the profit of any scheme containing the hall will be further reduced by the profit which would have been contributed by one shop and one flat. This lost profit must now be calculated, as follows:

Each shop contributes the following profit:
- zone A: 36 (sq.m. area) × 2189.4 (£/sq.m. profit)
  = £78,818
- zone B: 36 (sq.m. area) × 939.4 (£/sq.m. profit)
  = £33,818
- zone C: 36 (sq.m. area) × 314.4 (£/sq.m. profit)
  = £11,318
- therefore total profit contribution for one shop
  = £78,818 + £33,818 + £11,318
  = £123,954

Each flat contributes the following profit:
- 60 (sq.m. area) × 185.2 (£/sq.m. profit)
  = £11,112

So the total economic disadvantage caused to the scheme by including the community hall is the total cost of the hall itself (£150,000) plus the loss of profit from the shop (£123,954) and the flat (£11,112) which it replaces.

This totals £285,066; so to make the scheme viable, we shall have to build enough extra offices to generate this £285,066 extra profit. The necessary area of offices can now be calculated as follows.

- extra profit generated per sq.m. of offices
  = £365.9
- therefore extra area of offices needed to generate £285,066 profit
  = 285,066 ÷ 365.9 = 779 sq.m.

This produces a viable scheme of the form sketched below:

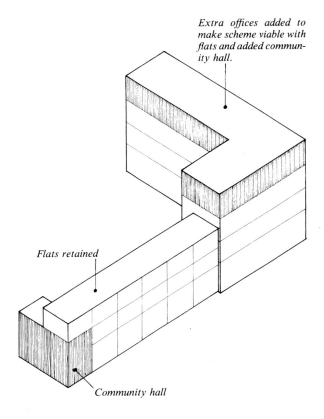

Extra offices added to make scheme viable with flats and added community hall.

Flats retained

Community hall

But adjusting the scheme to make the figures show an adequate profit does not automatically mean that the project will be attractive to the necessary development agencies. For example, we have located the community hall in the part of the scheme which we envisage being funded by a local builder/developer. We must check whether this would be acceptable to the developer or not, and whether the management problems it might cause could be successfully overcome. To make variety viable, therefore, we need to keep a constant focus on three different aspects of the project:

- the physical design (whose variety we are trying to maximise).
- the financial balance sheet (which indicates whether such a scheme could potentially be developed).
- the management requirements of the development agencies involved.

To achieve real variety, all three aspects *must* be developed together.

The case of social housing

The economic consideration of a social housing project will be rather different: in most countries, the assessment of the financial viability of such projects is unrelated to projected rental income. In Britain, for example, consideration at project level is primarily in terms of development costs.

Local government in Britain now rarely purchases land for housing, though some 'general family' housing, as well as accommodation for a wide variety of 'special needs' groups, is provided by registered housing associations. These associations are funded either by local authorities or more usually by the Housing Corporation, which is a central government agency[15].

Local Authorities themselves are now required to demonstrate a cost/value comparison for their schemes, based on estimated cost and projected value for the scheme. Many projects, however, still proceed with costs well in excess of valuation.

Proposals for all housing association projects are considered against a scale of *Total Indicative Costs* (TICs). Scales are regionally weighted and regularly uplifted to take account of inflation. It is intended that TICs should be *indicative* costs, rather than cost *limits*, and the funding authority may agree to a level of scheme costs at tender stage which is well in excess of the appropriate scale figure, in recognition of particular development difficulties, whilst having regard to the ill-defined criterion of 'value for money'.

TICs include land acquisition costs, construction costs and professional fees. The new-build TIC matrices[16] relate storey height and design occupancy - but not floor area - to cost. Rehabilitation matrices relate sizes of dwellings to cost.

The funding authority can only agree to the purchase of land at a price within the valuation set by the District Valuer, who is a local government official. To test the viability of a scheme, it is neccessary to estimate construction costs and professional fees in order to establish whether the residue of the TIC figure represents a realistic acquisition cost, as illustrated in the example below.

Example

A housing association is considering developing a project in outer London (TIC region). The vendor requires the District Valuer's agreed valuation of £100,000 as the price for the site, on which it is proposed to build six no. five-person (design occupancy) two-storey family houses, and six no. two-person flats in a three-storey block.

The TIC for a five person two-storey house in outer London is (say)£34,900, and for a two person flat in a three-storey block it is £26,000.

Construction costs for the project, including external works and car-parking, are estimated at £235,000. The architect's fees are to be 6% of the construction costs, the quantity surveyors require 2% and the structural engineers 1%.

The total government funding (TIC) available amounts to:

$$6 \times £34,900 = £209,400$$
$$6 \times £26,000 = £156,000$$

$$\text{Total} = £365,400$$

Total estimated construction costs and associated professional fees:

Construction	£235,000
6% architect's fee	£14,100
2% QS fee	£4,700
1% engineer's fee	£2,350
Total	£256,150

Residue available for land acquisition (land and legal fees):

TIC	£365,400
Estimated development costs	£256,150
Total	£109,250

Thus the residue of the TIC will cover costs, together with legal fees of $\frac{1}{2}$ - 1%, so the project is viable in financial terms.

In practice, provided the vendor is prepared to sell for housing association development, the District Valuer's valuation is usually sufficiently conservative to permit development. High value land, however, can rarely be used for housing association development; though in recent years packages have been put together on relatively prestigious sites, where it has been possible to include elements of housing association rental housing with private and shared equity housing, as well as commercial development .

When all else fails

Sooner or later, if the figures cannot be made to work out, there will be a temptation to alter the assumptions on which they are based: to convince ourselves that the scheme has such environmental quality that higher rents and lower yields could be achieved, or that costs could be cut by clever design, and so on. But there is no point in using unrealistic figures just to avoid facing facts: in the end, all this will achieve is a lot of abortive work.

However, now that the schedule of accommodation is beginning to firm up, it is certainly appropriate to check the assumptions you made at the beginning:

- check rents and yields with the estate agent, in the light of the design decisions already made. The value of the scheme will be particularly sensitive to changes in *yield*: a small decrease here will generate a lot of extra value.
- it is not worth spending too long thinking about construction costs, except for making a note of the obvious fact that if viability is difficult to achieve, then the scheme will have to be cheap. The QS will not be able to help you much further at this stage, because you do not have enough detailed information about the physical design to make detailed costings possible.
- finally, this is the stage at which to argue about acquisition costs. As a first step, you will have to work out the price you *could* afford to pay for the land to make the scheme viable.

Maximum acquisition cost

Let us return to our original (unviable) example, and call the maximum viable acquisition cost 'x'. Then the total cost of the project, as explained in Design Sheet 2.5, is obtained by adding the following components:
- acquisition cost (x in our example)
- construction cost (£1,611,360)
- fees (£241,704)
- short-term funding, calculated as follows:
  - Short term funding = average borrowing × years construction period × %compound interest rate ÷ 100
  - average borrowing = acquisition cost + (construction cost ÷ 2) + (fees × 2 ÷ 3)
  - = x + (£1,611,360 ÷ 2) + (£2,141,704 × 2 ÷ 3)
  - = x + £805,680 + £161,136
  - = x + £966,816

Therefore short-term funding
= (x + £966,816) × 2 × 0.1344 (from compound interest tables)
= 0.2688x + £259,880
Therefore total cost
= x + £1,611,360 + £241,704 + 0.2688x + £259,880
= 1.2688x + £2,112,944
and profit (at 20% of total cost)
= 0.254x + £422,588
Total value of project (see Design Sheet 2.4) = £3,324,000
As before, no external subsidies are available. For the project to be feasible, therefore,

(value + subsidy) > or = (cost + profit)

therefore
£3,324,000 + £0 = (1.2688x + £2,112,944) + (0.254x + £422,588)
therefore 1.5228x = £788,468
therefore x = £517,775 (say £515,000). So if the land can be acquired for £515,000 rather than our original estimate of £650,000, then the project will be viable.

By this stage in the design process, we have made decisions about a feasible schedule of accommodation for the project, generating as much variety as possible; and we have related the various uses together to minimise negative interaction between them.

Later stages of design, covered in the following chapters, will concentrate on making responsive physical form; but it is essential that the design is checked for feasibility at every stage in its development. From now on, physical and financial design should go hand in hand.

Chapter 3: Legibility

Introduction

So far, we have discussed how to achieve greater permeability and variety. But people can only take advantage of the choice which those qualities offer if they can grasp the place's layout, and what goes on there. *Legibility* - the quality which makes a place graspable - is the next topic to explore.

Different levels of legibility

Legibility is important at two levels: *physical form* and *activity patterns*.

Places may be read at either level separately. For example, it is possible to develop a clear sense of the physical form of a place, perhaps enjoying it only at an aesthetic level. Equally, patterns of use may be grasped without much concern with form. But to use a place's potential to the full, awareness of physical form and patterns of use must complement one another[1]. This is particularly important to the outsider, who needs to grasp the place quickly.

Why is legibility a problem?

The legibility of both form and use is reduced in modern environments. This is easily seen by comparing the traditional city with its modern counterpart.

Legibility and the traditional city

Before the twentieth century, cities worked well in terms of legibility. Places that *looked* important *were* important, and places of public relevance could easily be identified. This was true of outdoor spaces and buildings alike.

The biggest open spaces were related to the most important public facilities:

Delft, Netherlands

The buildings which stood out from the rest were those of greatest public relevance:

Where privacy and security permitted, many buildings allowed the passer-by to see the activities inside:

The modern city

The modern city is legible only in the sense that 'buildings cannot lie': large office blocks, owned by pension funds and insurance companies, occupy key city centre positions; expressing the power of big financial institutions. But these bureaucratic enclaves - irrelevant to how most people use the city - visually overwhelm publicly-relevant places and facilities, confusing important activity patterns.

This confusion is made worse because important public buildings and publicly-irrelevant private ones often look alike.

Separating pedestrians from vehicles

Finally, the desire to separate pedestrian and vehicular routes - remarked on in Chapter 1 - makes both central and suburban areas far less legible for people on foot. In the suburbs, pedestrians are all-too-often dragged confusingly round the private backs of the houses, between mute fences and planted privacy screens. Offering little to remember, such places are hard to grasp.

In the town centre, pedestrians are expected to follow ill-defined paths, sometimes underground, sordid and alienating, threaded tortuously up and down through the gaps between vehicular roads.

Achieving greater legibility

Legibility is being continuously eroded, so increasing it is an important objective, affecting the design of both physical layouts and patterns of use. In practice, it is easiest to start by considering layout. We shall return to activities later in this chapter.

Legible physical layouts

The point of a legible layout is that people are able to form clear, accurate images of it. Note that it is the *user*, rather than the designer, who forms the image: the designer merely arranges the physical layout itself.

Various researchers have explored the content of these images; using such techniques as interviews, asking directions to places, and getting people to draw maps from memory, like the one below.

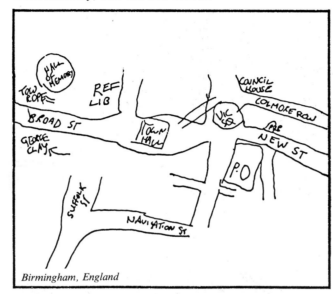

Birmingham, England

Overlapping images

Analysis of data like these reveals a considerable overlap between different people's images of a given environment, enabling a shared image to be mapped. An example - 'the Boston everyone knows' - is shown below[2].

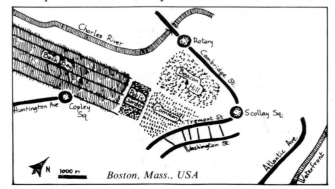

Boston, Mass., USA

Key physical elements

Certain sorts of physical features play a key role in the content of these shared images. Kevin Lynch - the American planner who pioneered studies of this topic in the 1960s - has suggested that these features can be grouped into five key elements, as illustrated below[3].

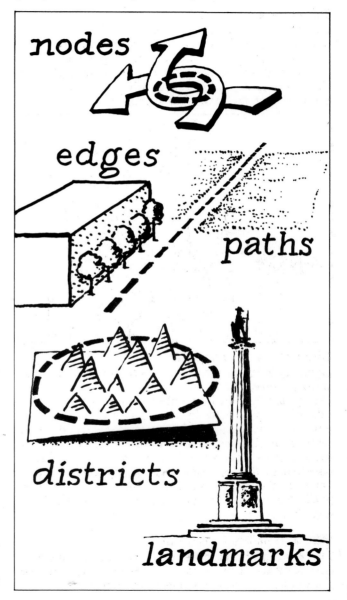

nodes

edges

paths

districts

landmarks

Paths

Paths are amongst the most significant of these elements. They are channels of movement - alleys, streets, motorways, railways and the like - and many people include them as the most important features in their images of the city.

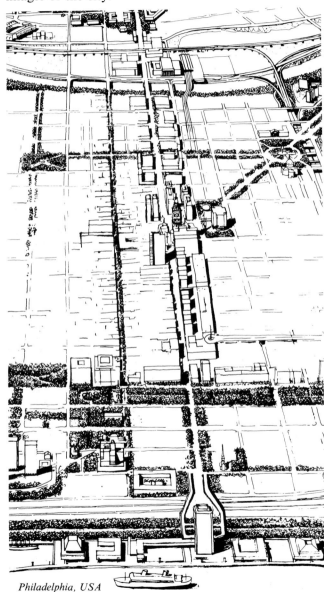

Philadelphia, USA

Nodes

Nodes are focal places, such as junctions of paths: examples extend from roundabouts to market squares.

Cambridge Circus, London, England: Cruikshank.

Landmarks

In contrast to nodes, which can be entered, landmarks are point references which most people experience from outside.

Boston, Mass., USA

Edges

Edges are linear elements which are either not used as paths, or which are usually seen from positions where their path nature is obscured. Ralph Erskine's Byker wall is an example of the first type, whilst the second includes elements like rivers, railway viaducts and elevated motorways.

Byker,   Newcastle, England

Districts

Paths, nodes, landmarks and edges constitute the skeleton of the urban image, which is fleshed out with areas of less strongly differentiated urban fabric. The distinction between skeleton and flesh comes over strongly in the eighteenth century engraving of Paris shown below. The flesh itself is organised into *districts*: medium-to-large sections of the city, recognisable as having some particular identifying character. A composite residents' map of districts in part of Boston is shown on the right[4].

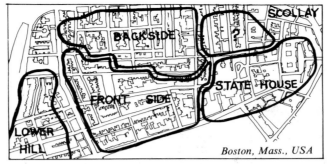

Boston, Mass., USA

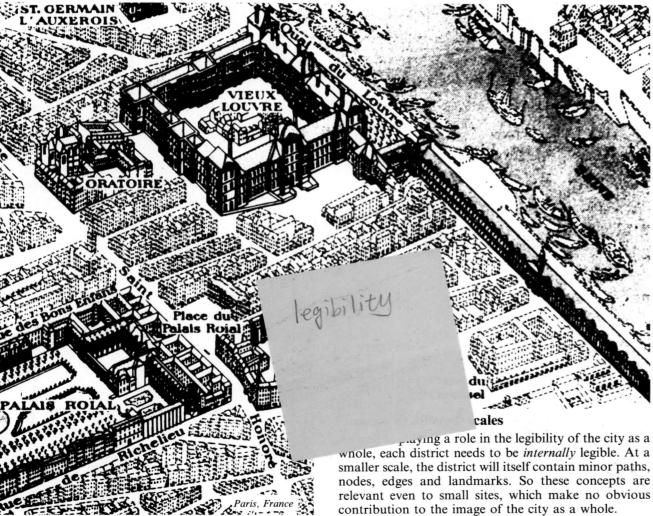

Paris, France

Using the elements

Firstly, though the elements are themselves *abstractions* rather than concrete forms, a designer aware of their importance for legibility is helped to focus on the *kinds* of physical forms worth taking as models for legible new layouts.

Secondly, thinking in terms of these elements helps designers analyse the key image-forming features - both actual and potential - in their projects' existing surroundings: research suggests that familiarity with these concepts enables reasonably accurate prediction of the features of a place which are likely to form key parts of its users' images[5]. Design Sheet 3.1 explores how this can be done in practice.

This kind of analysis is all the more useful when the designer can get a wider public involved in it. Ways of doing this are discussed in Design Sheet 3.2.

Combining new and existing elements

The first step in design is to develop the project to make more legible the area of which it forms a part, by relating the new design to existing elements on the site and in its surroundings. Because these cannot be moved, they must be taken as fixes for developing the design. The implications of existing paths, nodes, edges and landmarks are discussed in Design Sheet 3.3; whilst those of existing districts are covered in Sheets 3.4 and 3.5.

Having related the scheme to significant existing elements, we turn our attention to the new elements within the project itself. It is sensible to start by considering paths: as we have seen, they are often the most important features in people's images of places; and in any case our starting point for design development is the permeable path system worked out in Chapter 1.

... playing a role in the legibility of the city as a whole, each district needs to be *internally* legible. At a smaller scale, the district will itself contain minor paths, nodes, edges and landmarks. So these concepts are relevant even to small sites, which make no obvious contribution to the image of the city as a whole.

Reinforcing paths

There are two objectives to be achieved in reinforcing path legibility:
- to give each path a strong character, easily distinguished by users
- to bring out the relative importance of each path, as decided in Design Sheet 1.2.

The design implications of these objectives are discussed in Design Sheet 3.6.

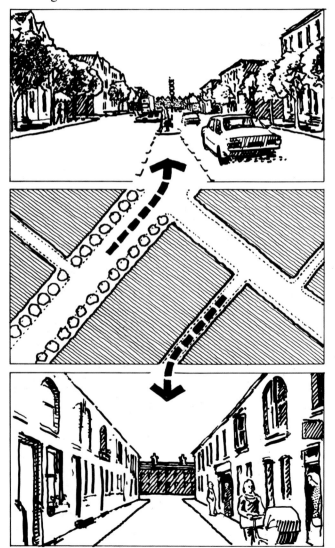

Reinforcing nodes

By this stage, the positions of all the nodes are fixed. The next step is to decide how far the legibility of each should be reinforced. This depends on two main factors:
- the functional roles of the linking streets (as discussed in Design Sheet 1.3).
- the level of public relevance of the activities in the adjacent buildings.

The design implications of both factors are discussed in Design Sheet 3.7.

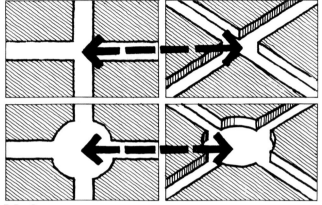

Marker sequences

By now, the paths have already been differentiated from one another by differences of width and enclosure, and the nodes have been reinforced as markers within the path system. But sometimes additional intermediate markers are needed, to remind users of their position along the path concerned, and to give a sense of getting somewhere. These final elements for reinforcing the project's legibility are discussed in Design Sheet 3.8.

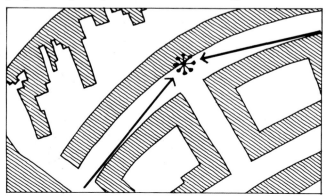

Design implications

How to achieve legibility

1. **Take the street/block layout and schedule of accommodation from Chapters 1 and 2 as the starting point for developing legibility.**

2. **Assess the existing legibility potential of the site and its surroundings (Design Sheet 3.1)**

3. **Check this assessment against the views of a wider public, as far as resources permit (Design Sheet 3.2)**

4. **Adjust the project's street/block layout to make the best use of the legibility potential of existing elements on and around the site (Design Sheet 3.3)**

5. **Assess which district the site belongs to, and the consequent design implications (Design Sheet 3.4)**

6. **Where the project's district has strong path themes, develop an appropriate vocabulary of building heights and street widths for the new design (Design Sheet 3.5)**

7. **Check that path enclosure is adequate for legibility (Design Sheet 3.6)**

8. **Reinforce legibility of nodes within the scheme, according to their relative importance (Design Sheet 3.7)**

9. **Introduce intermediate markers into the path system if necessary (Design Sheet 3.8)**

3.1: Legibility analysis

Since nowhere is totally illegible, start by finding out the existing potential of the site and its surroundings. Look for any existing activities and forms which could be used to make the place more legible, and record them - and how they might be used - on a plan.

You should cover any nearby areas which can be seen from the site, as well as the area of the site itself. Since the new design should contribute to the legibility of its surroundings as well as being legible in itself, pay special attention to any parts of the site which can be seen from anywhere outside it.

It is often helpful to use Lynch's checklist of elements - paths, nodes, landmarks, edges and districts - to stimulate this analysis. Typical factors to look for include the following:

- Paths:
  record any routes which adjoin or cross your site, noting their relative intensity of use, as discussed in Design sheet 1.3.
- Nodes:
  note any place where paths meet; recording the relative importance of each path, and the public relevance of any associated buildings.
- Landmarks:
  record any publicly-relevant activities, either in buildings or in outdoor spaces.
- Edges:
  record any distinct limits to areas with different patterns of use or visual character.
  record any strong linear barriers.
- Districts:
  record areas with different patterns of use.
  record areas with different visual characters, and decide what makes the differences; overall building forms, materials or details.

Do not let this list become a straightjacket: it is quite wrong to assume that every area contains each type of element in the list.

The drawing on the right shows the use of this approach to analyse the legibility potential of a redevelopment site at Oxford Railway Station. This records one designer's opinions about legibility, both potential and actual. The next step is to check, as far as resources allow, whether the elements recorded in the analysis actually make the place legible to its users. Ways of doing this are discussed in Design Sheet 3.2.

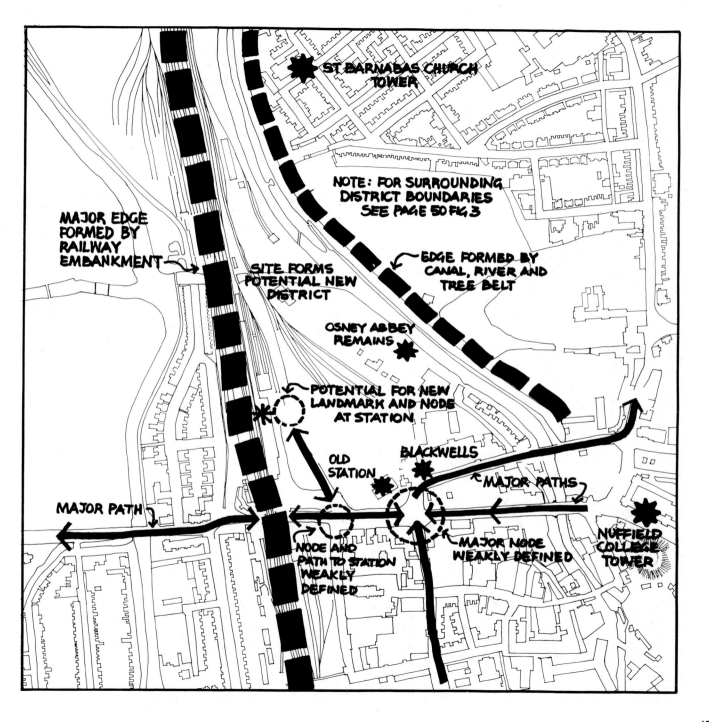

3.2: Legibility and the user

It is important to check your own assessment of the site's legibility against the views of a wider public, as far as resources permit.

Who should you approach?

To make sure the exercise will be useful, the people you interview must be carefully selected:

- Choose people who regularly use your site, or its immediate surroundings. What you need is *detailed* information: general views about the town as a whole are of little use for site-specific design.
- Approach as wide a range of the site's users as possible. Try to balance out sexes and age groups in your 'sample'.
- Whichever technique of enquiry you use, aim to explore the views of about 20-30 people.

What are you trying to investigate?

The purpose of your investigation is to test and modify your ideas about the site's legibility, as developed in Design sheet 3.1. So begin by listing all the ideas to be tested, from the annotated 'problem and potential' and 'response' plans in Design Sheet 3.1. Then consider which techniques of enquiry might best elicit the information you need.

Local education resources

Mental mapping exercises are now quite common in secondary environmental education: you may find that a local school or college has already done work on your area, and has a thesis covering your site in detail. If not, they might help if the work fits into their curriculum. This would probably give a more rigorous coverage than you could achieve yourself.

Local workplaces, cafes and pubs

Ask people to help you during a break: this is the most likely way to get people drawing their own maps. It is often useful to provide a standard xeroxed sheet showing some important nearby features. You are not after complete recall, and should stress that accuracy is not the answer. You are after how people see the area, not how it is: the end product is likely to be something like the example shown below. Aim for 20 to 30 maps. summarising oft-mentioned features and overlapping boundaries.

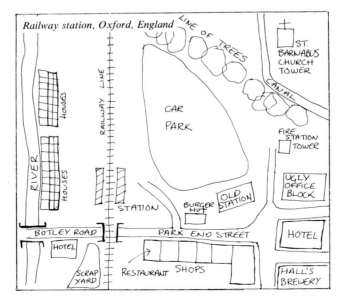

Railway station, Oxford, England

Doorstep interviews

These are useful for tapping the views of local residents. Rather than request maps, it is often better to stimulate discussion through the use of photographs, especially of the 'do you remember when this building stood over there?' variety. You can steer the conversation to cover your own hypotheses about the site, but be careful not to lead the answers: always remember that people on their own patch are the experts.

Street corners

Ask questions about the place, with one or two photographs to stimulate conversation. You will be lucky to get answers from more than one in ten people, but in some situations this may be the only way of contacting regular site users. But remember that the police may be suspicious of people conducting public interviews: if you intend to use the 'street corner' approach it may save embarassment if you tell the local police station what you are doing in advance. Policemen are often mines of useful information anyway.

By this stage, you should have a clear idea about the potentials and problems of your site in legibility terms. The Design Sheets which follow show ways of using this information to develop your design in more detail.

3.3: Combining new and existing elements

The last two Design Sheets highlighted the legibility potential of existing elements on the site, or visible from it. Now use this information to develop the tentative layout worked out in Chapters 1 and 2, to make a legible scheme which will also contribute to the legibility of the area around it.

Legibility depends on the relationships *between* elements, even more than on the design of the elements themselves. The forms and positions of the existing elements noted in the legibility analysis are already fixed. So the only way they can be used in developing the legibility of the place is by relating the layout of the new scheme to *them* as legibly as possible.

The first step towards achieving this legible relationship is to take the tentative layout of streets, blocks and uses developed in the last two chapters, and superimpose it on the legibility analysis plan worked out in Design Sheet 3.1.

Without losing the practical block sizes and shapes developed in Design Sheet 1.4, adjust the positions of the street/blocks and the uses they house, to form the most legible relationships with the existing elements. At this stage, concentrate on putting the overall skeleton of the project together in the most legible way. Do not worry about the design of the individual parts of the scheme as yet.

Work through the existing elements as outlined below:

Existing paths and nodes
Make sure they are all defined by the edges of the blocks.

Existing landmarks
Allocate publicly relevant uses to landmark buildings, and adjust street alignments to make the most positive use of them. Various approaches are possible:
- focus streets on them.
- position new nodes adjoining them, thus using each landmark as a focus for several streets.
- incorporate them into streets as intermediate markers.
These ideas are illustrated on the right.

Existing edges and districts
These need not be considered yet. Their implications will either be too detailed to be relevant at this stage or - in the case of linear barriers - they will impose themselves on the design whether you want them to or not.

After working through this process, the project's layout will have been adjusted to make the best use of the legibility potential of the existing elements on and around the site. The next step is to develop the legibility of the *new* elements. This is discussed in the Design Sheets which follow.

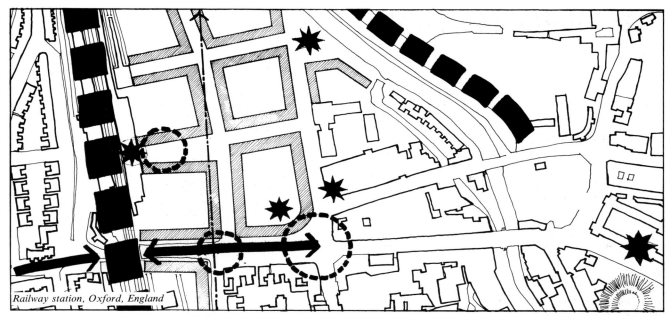

Railway station, Oxford, England

1. Layout before combining new and existing elements

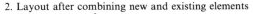
2. Layout after combining new and existing elements

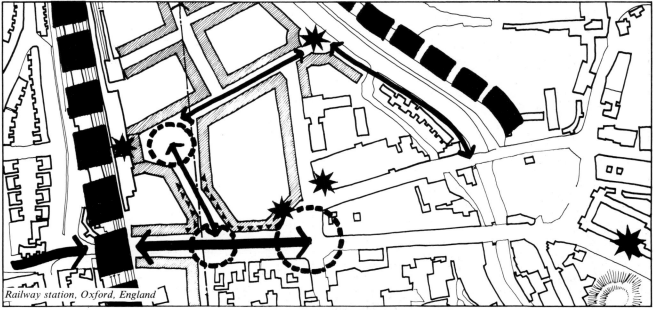

Railway station, Oxford, England

3.4: District location

Design Sheets 3.1 and 3.2 noted whether any of the legibility potential of the site or its surroundings was due to the area being perceived as organised into districts. This analysis must now be taken a stage further; firstly to decide *which* district the site itself belongs to - if this is not already obvious - and secondly to investigate the implications of this district setting for the design of the new scheme.

Which district is the project in?

This may be obvious (1). But the site may lie at the junction of districts, potentially linked to either (2). Or a large site might be regarded as a *new* district (3).

If it is not obvious which district the project belongs to, consider the following:
- district character often depends on consistent patterns of uses. Is this the case with any of the districts to which the project might relate? Do the uses in the scheme make it easier to integrate with one district rather than another?
- district character often depends on the repetition of typical building themes: heights, frontages, materials, details and so forth. Do any of the design decisions which have so far been made fit more easily into one district than another?
- the various possible districts may differ in economic terms: some may be declining, others static or buoyant. If so, does this imply which district the project must seem a part of, in order to attract the variety of uses investigated in Design Sheet 2.1?

Having decided to which district the project belongs, the next step is to assess whether this has implications for designing the massing, street layout and public space enclosure with which this stage of design is concerned. This depends on whether such factors are important to the district's character. Where they are, Design Sheet 3.5 discusses how to use them in design. Where they are not, go straight on to Design Sheet 3.6

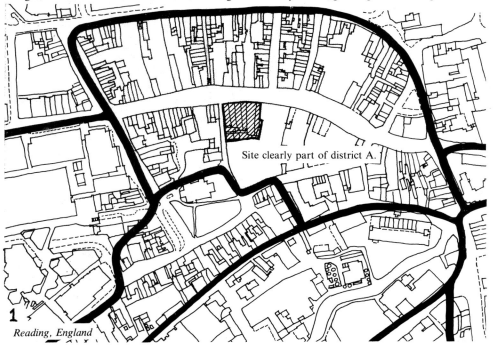

Site clearly part of district A.

Reading, England

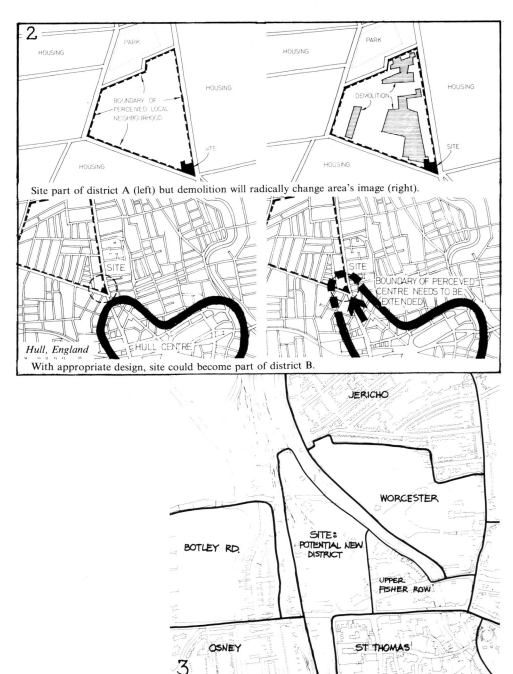

Site part of district A (left) but demolition will radically change area's image (right).

Hull, England

With appropriate design, site could become part of district B.

Oxford, England

3.5: Districts with strong path themes

If massing and street enclosure are important to the district's character, they must be investigated in more detail. Analyse these factors in the project's own district, and in those adjoining it. Note plan dimensions, and enclosure in both plan and section. The purpose of this exercise is to home-in on a range of dimensions which are typical of the project's own district, and clearly distinct from those in adjoining areas; thus reinforcing the differences which distinguish one district from another. The sketches on the right show how this might be done.

This process will suggest a vocabulary of street dimensions to support the existing district character. To maintain and develop the legibility of the city as a whole - which is helped by clearly distinguishable districts - this vocabulary should be used in the new project unless one of the following considerations applies:

- a district with a *very* distinct, homogenous character may contribute to the legibility of the city as a whole; but its homogeneity may make it illegible internally. In this case, ignoring the district's themes in the new design may increase the internal legibility of the district, without significantly weakening legibility at the city scale. But this depends on the *size* of the scheme: large projects which ignore district themes may erode district character altogether.
- more positively, if the scheme is of particular public relevance *within* a homogenous district, it may be appropriate to emphasise it by ignoring the district's themes, as in the example below:

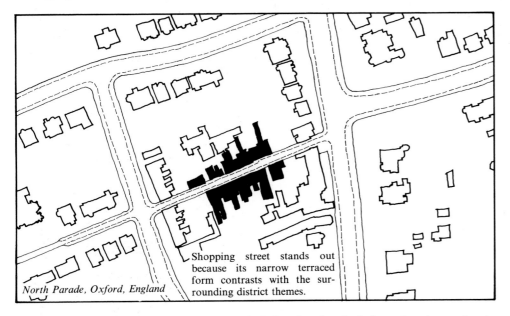

Shopping street stands out because its narrow terraced form contrasts with the surrounding district themes.

North Parade, Oxford, England

Finally, the district's path vocabulary should be checked against the criteria for path enclosure given in Design Sheet 3.6. These should only be overridden when there is no other way of sufficiently reinforcing the district character. So consider whether this reinforcement could be achieved with detailed design, as discussed in Chapter 5.

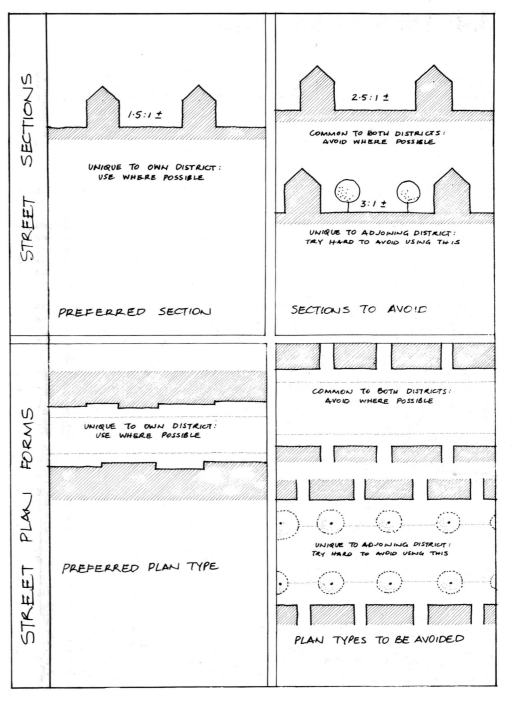

STREET SECTIONS

1.5:1 ±

UNIQUE TO OWN DISTRICT: USE WHERE POSSIBLE

PREFERRED SECTION

2.5:1 ±

COMMON TO BOTH DISTRICTS: AVOID WHERE POSSIBLE

3:1 ±

UNIQUE TO ADJOINING DISTRICT: TRY HARD TO AVOID USING THIS

SECTIONS TO AVOID

STREET PLAN FORMS

UNIQUE TO OWN DISTRICT: USE WHERE POSSIBLE

PREFERRED PLAN TYPE

COMMON TO BOTH DISTRICTS: AVOID WHERE POSSIBLE

UNIQUE TO ADJOINING DISTRICT: TRY HARD TO AVOID USING THIS

PLAN TYPES TO BE AVOIDED

3.6: Path enclosure

There are two objectives to be achieved in designing path enclosure:

- to give each path a *strong character*, easily distinguished by users
- to bring out the relative *functional importance* of each path, as decided in Design Sheet 1.3.

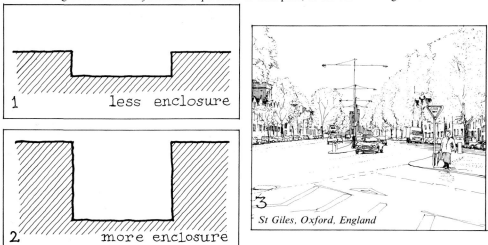

A path's legibility is crucially affected by its enclosure in plan and section. Height/width ratios of less than 1:3 seem weakly enclosed (1,2), so avoid them where possible. Where this is difficult, enclosure can be increased by planting (3).

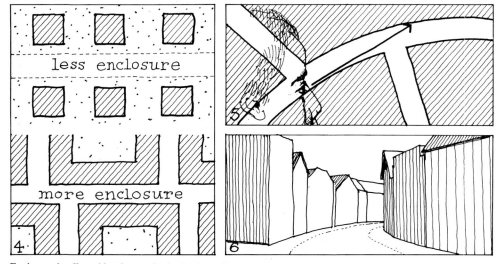

Enclosure is affected by the continuity in plan of the enclosing elements (4), and by the form of the path as a whole (5,6).

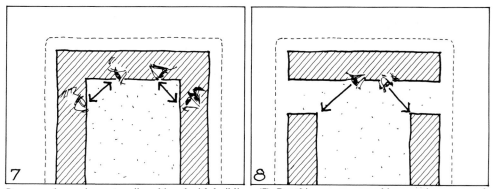

Strong enclosure is most easily achieved with buildings (7). But this may cause problems at the corners of blocks: particularly with housing, privacy may be destroyed by overlooking between windows of adjacent dwellings at the internal angle (8). This is often solved by leaving a gap at the corner of the block

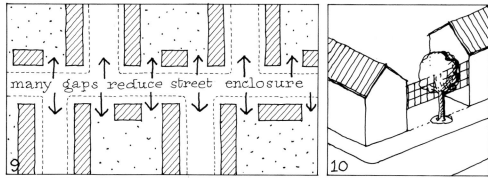

With small blocks, this produces a large number of gaps, and consequently reduces street enclosure (9). These gaps can be closed by walls, trellises or trees (10). But these contribute no activity to the street, which becomes correspondingly less memorable. L-shaped corner houses, partly single-aspect to avoid overlooking, can also be used.

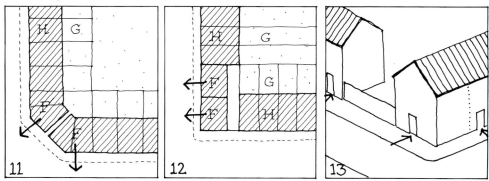

Corners can also be closed with single-aspect flats (11, 12). Where a gap is left, it will help if the corner house has its entrance at the side, to reduce the length of blank wall on the 'gap' street (13).

Finally, check that no paths are confusingly similar. Problems here can be solved at a detailed architectural level (as discussed in Chapter 5) and by giving the paths concerned perceptually different markers, as discussed in Design Sheet 3.8.

3.7: Nodes

By this stage, the positions of all the junctions and the uses in the buildings around them have been fixed. The next step is to decide which junctions need special reinforcement, and how this should be done.

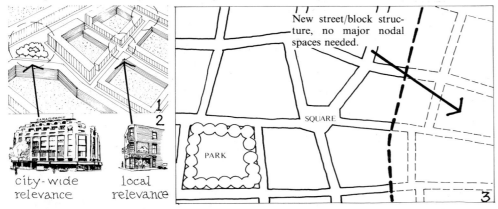

All junctions are potential nodes, but they should not all be given equal significance. The appropriate degree of emphasis for each node depends on three main factors:
- the functional roles of the streets forming the junctions (as discussed in Design Sheet 1.3): the more important the functional role, the greater the spatial emphasis required to maintain the congruence between legibility of use and legibility of form (1).
- the activities in the adjacent buildings: for the same reason, the more publicly-relevant these are, the greater the spatial emphasis required (2).
- the expectations set by other nodes within the district concerned: these establish a vocabulary from major to minor nodes, within which the new one should fit if its relative importance is to be grasped easily by its users (3).

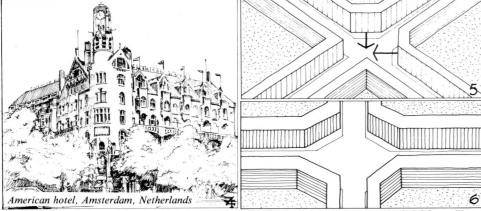

Design sheets 3.1 and 3.2 identified existing nodes and the places where new ones are needed. Most will not require major spatial reinforcement. But their legibility can be increased, if necessary, by distinctive corner buildings (4). Splayed corners can help at crossroads because they focus the buildings on the centre of the space (5). Splayed corners also give the junction a greater sense of enclosure because they start to form a concave shape (6). But be careful not to increase the height/width ratio above 1:3.

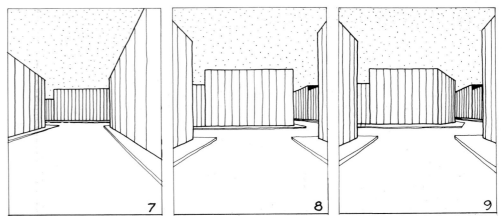

Offsetting the junctions also increases the sense of enclosure: as you approach the node, there is a building immediately ahead closing the view (7). But there is the danger of reducing visual permeability, so keep the offset as small as possible (8). Splayed corners and set-backs can help by deflecting the eye towards the next path, as well as increasing the concavity of the space still further (9).

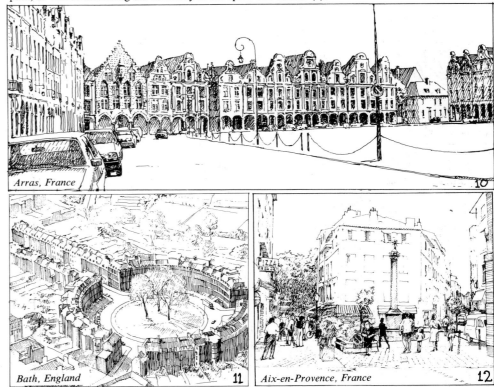

Arras, France

Bath, England

Aix-en-Provence, France

The forming of a concave nodal space is one of the most emphatic ways of increasing the legibility of the junction. The urban square or circus is one of the strongest examples of this. It can be used at a variety of scales (10, 11, 12).

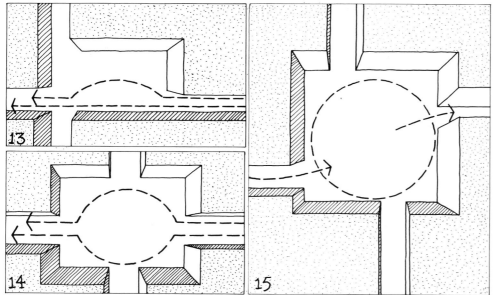

Where nodes are large, there are more possible entrance positions. If the wall defining the entrance path continues uninterrupted, to form the wall of the node (13) then the node itself may read as a mere widening of the path. With entrances located away from its corners (14) the node seems more distinct from the paths leading into it. This effect is strengthened if it is impossible to see straight through from entrance to exit (15). But here the increase in spatial definition must be weighed against the possible loss of visual permeability.

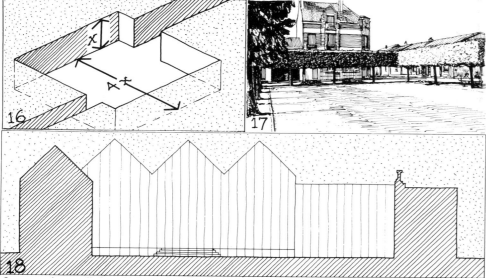

Larger nodes usually have higher ratios of enclosing wall to street opening on plan, but are more difficult to enclose in section. Because of greater plan enclosure, height/width ratios can be opened up to about 1:4 before the enclosure seems too weak (16). Effective width can be reduced by trees or walls (17), or height can be increased by roof pitches, balustrades or changes in ground level (18).

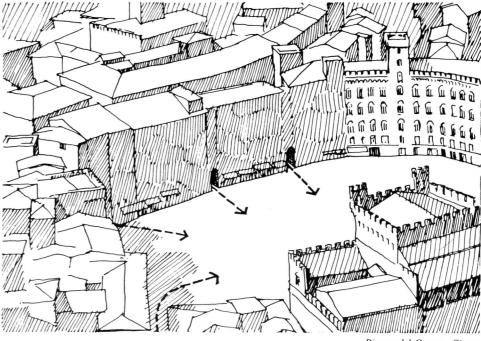

Piazza del Campo, Siena.

Though contrasting, both these examples use the ideas we have outlined:
-   compact planforms, with strong enclosure in section and plan.
-   entrances positioned to emphasise the node as an element distinct from its surroundings.
-   entrances designed for minimum interruption of the node's enclosing surfaces.

In both the places shown, designers have allowed the forms of buildings and planting to be dictated by the need for legible urban places.

David Walker: Roundabout with circular hedge, Art into Landscape competition, 1977.

3.8: Marker sequences

Design Sheet 3.6 distinguished paths from one another by differences of width and enclosure, while Design Sheet 3.7 discussed nodes, which act as markers to help users locate themselves within the path system as a whole. But in some situations, further intermediate markers are needed to show users where they are along the path concerned, and to give a sense of getting somewhere. When streets are straight and junctions are frequent, the path is unlikely to need other markers for its *own* legibility. But it may contain publicly-relevant uses, treated as landmarks to make their own roles legible, and thereby forming extra street markers. If these are sited at junctions, they will benefit from their visual exposure to several routes, and will in turn contribute to the legibility of the junction, supporting visual permeability still further.

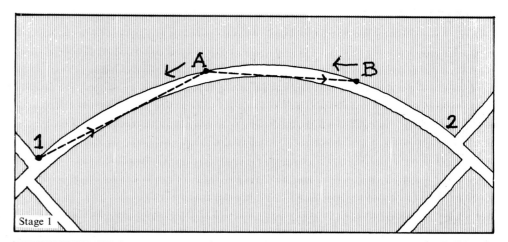

Stage 1

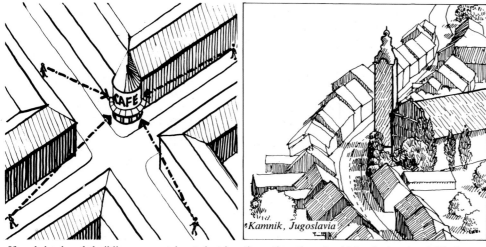

Kamnik, Jugoslavia

If such landmark buildings are not located at junctions, then they should be sited to be visible from a distance. They will need to project forward on plan, or upwards or downwards in elevation, relative to the adjacent street fronts.

When siting markers in a curved street, slightly different considerations apply. Junctions may not be visible from each other, so they may need to be supplemented by extra markers to achieve a legible street. The widest spacing of these markers can be decided as shown in the drawing on the right:

- starting from junction 1, draw the longest possible sight line towards junction 2 until it hits the building line at A.
- draw an arrow pointing back from A towards 1 as shown, to indicate that a marker is needed somewhere between 1 and A.
- from A, draw the longest possible sight line onwards towards the second junction, until it hits the building line again, at B. Draw an arrow back towards A, to show that a marker is needed between B and A.
- continue drawing sight lines, and arrows, until the second junction (2) is reached.
- then repeat the procedure working back from the second junction (2) towards the first, plotting points X (with arrow pointing back towards 2) and Y (with arrow back towards X).

A continuous visual chain of markers will be formed, using the minimum number of markers, if one is placed anywhere in each of the zones where the arrows point towards each other (Y→ ←A, X→ ←B in the example).

Now all the markers are positioned, their individual design must be considered. To maintain congruence between legibility of use and legibility of form, try to associate marker features with publicly-relevant activities. Whether this is possible or not, ensure that the markers stand out visually from their surroundings: ways of achieving this will be discussed later, in Design Sheet 5.3.

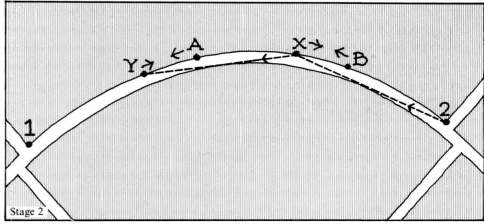

Stage 2

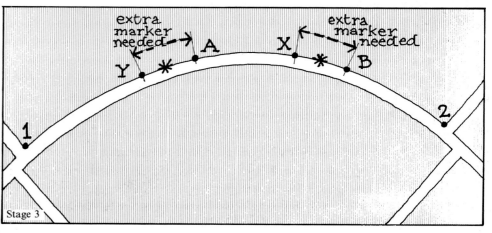

Stage 3

Chapter 4: Robustness

Introduction

Places which can be used for many different purposes offer their users more choice than places whose design limits them to a single fixed use. Environments which offer this choice have a quality we call *robustness*.

Why is robustness a problem?

Whether we like it or not, the ultimate power of deciding how a place should be designed lies in the hands of whoever pays for it: the *patron*. Patronage is almost never controlled by the direct users of the place, who therefore have little say in its design.

Patrons are not usually interested in promoting user choice, because they are each concerned only with some particular aspect of a user's life: the user as rent payer, or office-worker, or car driver and so on. Because the particular activities defined by the patron get most of the designer's attention, projects are usually designed rigidly around them by tailoring the pattern of spaces so that the desired pattern of activities can take place as efficiently as possible, without interfering with each other[1]

Problems inside buildings

Inside buildings, this leads to a tendency for designers to provide *specialised* spaces for the different activities. This specialisation, to serve the patron's interests, often makes it more difficult for *other* activities to take place: this reinforces the effects of the patron's lack of interest in user choice.

So much for the ideal work-sequence...

Problems in public outdoor space

In public outdoor space designers tend to employ the same approach; thinking in terms of specialised spaces for different activities, separated off from one another. But activities in public space are *public* activities: they rarely need to be separated from one another for reasons of privacy. Indeed, in public space, it is the *activities themselves* which act as the most important supports for other activities: people come there to experience other people. So if public space is chopped up into separate compartments for separate activities, most of its robustness is removed.

Housing · Sun and shade · Offices · Retail · Cafés · Ball games

Treviso, Italy

What can designers do about this?

Designers cannot change the way patronage works, but they do not need to make the problem *worse* by the way they design. Given that patrons have the power, and will use it primarily to further their *own* interests, there is still nearly always some room for manoeuvre in designing for robustness, even when patrons are not prepared to pay extra for it.

Robustness and normal costs

Robustness can be increased, *within* normal cost limits, merely by careful design of the things which would have to be included anyway. Since it costs no more, this approach to robustness should always be pushed as far as possible.

Bedford Square, London, England: 150-year-old houses now contain offices, embassies and an architecture school.

Where shall we begin to design?

Robustness is equally important indoors and out, but its design implications for buildings are different from those for outdoor places. Particularly in urban situations, the activities in outdoor places are strongly influenced by what goes on in the buildings round their edges. So we shall take the buildings as our starting point, working outwards from them into the adjacent outdoor places.

In the context of buildings, it is useful to distinguish between *large-scale* and *small-scale* robustness.

Large-scale robustness

Large-scale robustness concerns the ability of the building *as a whole*, or large parts of it, to be changed in use.

Taking advantage of robustness at this scale usually involves resources which are not easily available to most people. But, indirectly, large-scale robustness can offer more choice to ordinary users in the long run. As buildings grow older, and move down-market, it becomes *financially* feasible for them to accommodate a greater range of uses, as we saw in Chapter 2. Large-scale robustness ensures that this is also physically feasible, and therefore makes it easier for the variety of uses in the area to increase.

Small-scale robustness

Small-scale robustness concerns the ability of *particular spaces* within the building to be used in a wide range of ways.

This is the scale of robustness most relevant to the majority of ordinary users. It is important because it has a direct effect on the day-to-day choices most people can make.

Design implications at different levels

Because it is concerned with major changes of use, large-scale robustness has implications for the *overall* design of the building, which need to be considered early on. Small-scale robustness involves design decisions of a more detailed kind: though they are of critical importance to users, they can safely be left till later. We shall therefore begin by designing for large-scale robustness.

Designing for large-scale robustness

We cannot predict the likely changes in use which might occur during the expected life of a building: even in the short run, predictions of this kind are notoriously unreliable. It is more practical to learn from buildings which *have* successfully coped with changing uses[2]. But the lessons to be learned are different for family houses than for other building types.

Family houses

The most important factor affecting the large-scale robustness of a given house design is the floor area it provides. So robustness is supported by opportunities for enlarging the house as a whole. This has many design implications, which are explored in Design Sheet 4.1.

Other building types

Experience suggests that there are three key factors which support long-term robustness,[3]:
- building depth
- access
- height

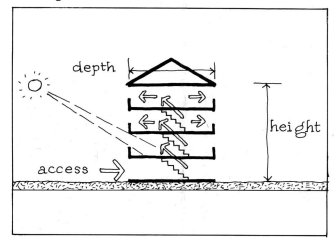

Building depth

The vast majority of building uses require natural light and ventilation. Buildings which are too deep for this cannot easily change in use.

Access

All building uses need *some* links to the outside world. So the number of access points is a key factor governing how easily a building can adapt to a variety of uses.

Building height

The importance of access also affects building height: in a tall building, the upper floors have restricted links to the outside, and are therefore less suitable for a wide range of uses.

Preferred configuration

Between them, these three factors define a preferred building configuration for achieving large-scale robustness:
- shallow in plan
- many points of access
- limited height

Of course, not *all* buildings can take this form: an international swimming pool, for example, wouldn't work with such a configuration. But only a small proportion of the building stock consists of buildings with such specialised requirements. And even those usually have less specialised parts which could use our preferred arrangement.

The design implications of these topics are covered in Design Sheet 4.2.

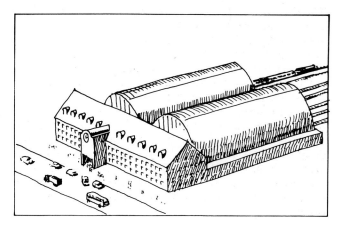

Internal organisation

As a starting point for designing the internal organisation, we have to accept the *patron's* view of how the building will be used. We must locate activities together, and give them spatial enclosure, in a way which patrons will accept as efficient for achieving their own objectives. We shall not discuss how to do this: it is already part of our current design tradition.

But this still leaves some freedom, because there are usually several design alternatives for achieving an efficient plan. If we are to use this freedom to build in as much additional robustness as we can, we must take some *extra* factors into account.

In most buildings, the various parts have different potentials for contributing to robustness. Two sorts of areas need special attention:
- hard/soft
- active/passive

Hard and soft areas

Most buildings contain spaces which house shared facilities such as staircases, lifts and vertical service ducts. Usually these spaces are 'hard': they are least likely to change their functions during the building's life. These hard zones must be positioned where they will not restrict the use of the remaining space. Ways of achieving this are covered in Design Sheet 4.4.

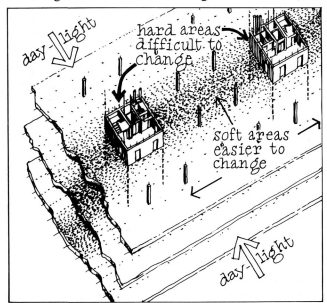

Active and passive areas

To an important extent, the potential for robust *outdoor* spaces depends on what goes on in the parts of the building immediately next to them. This must be taken into account when planning the buildings themselves.

Some activities within the building may benefit from being able to extend outwards into adjacent public outdoor space. When this occurs, they will contribute to the activity in the public space itself.

Other indoor activities may contribute to the level of outdoor activity: visual contact with them can make the place more interesting for spectators.

Any indoor area which can contribute to outdoor activity in either of these ways is called an *active* area. At this stage, we must decide which elements of our scheme have this active quality. So far as possible, we must ensure that the ground floor of the building, where it abuts public space, is occupied by these active areas. This is discussed in Design Sheet 4.3.

Small-scale robustness

Designing for small-scale robustness involves working at two levels:
- adjusting room sizes and shapes within the general spatial layout already decided.
- designing each room in detail.

Room size

Very small rooms can accommodate very few different activities, whilst very large ones can cater for a wide range. But above a certain size, as shown in the graph below, further increases begin to be less and less effective in accommodating more activities[4].

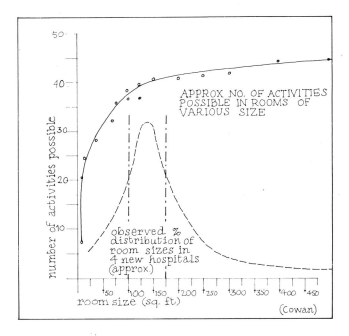

This means that there is an *optimum* room size: the 'best-buy' in terms of the number of activities which can be accommodated for a given floor area, offering the most choice for a given expenditure.

Room shape

Room shape also affects the number of different activities which can take place in a given area. Compact rooms are better than long thin ones in this regard, and are therefore more cost-effective in terms of the choices they provide.

Detailed room design

As well as the room's size and shape, its detailed design has an important impact on the number of different activities it can house. If carefully considered, factors like the positioning of doors, windows, socket outlets and radiators can contribute significant increases in robustness at no extra cost.

Robust room sizes, shapes and details are covered in Design Sheet 4.5.

Design for alteration

Finally, it is important to make it practicable to change as many as possible of our internal layout decisions during the life of the building. This is largely a matter of designing the physical fabric of the building so that it can easily be altered.

Outdoor spaces

Now we have decided the internal arrangement of the buildings, we can turn to the adjacent outdoor spaces, both public and private.

Private garden space

Outdoor space which is private, within the perimeter block, greatly increases the robustness of the surrounding buildings, particularly when these contain housing. Detailed garden design should be left to the users, but garden robustness is also affected by broader issues, which are discussed in Design Sheet 4.6.

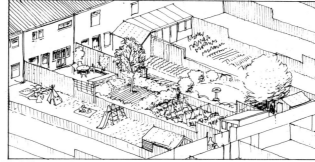

Public outdoor space

The design of public outdoor space is a complex matter. We begin by considering the edges of the space, because it is here that most activity takes place: for most people, in most places, the edge of the space is the space[5].

Piazza San Marco, Venice, Italy

Designing the edge of the space

We begin by capitalising on the active elements already located on the ground floors of the surrounding buildings. Ways of helping these to animate the adjacent edge of of the outdoor space are discussed in Design Sheet 4.7.

Veurne, Belgium

Having considered the activities at the edges, we can turn our attention to designing the main body of the space.

Designing within the space

The principle for supporting robustness is to design settings which, as far as possible, enable a variety of activities to co-exist in the public realm without inhibiting each other. This particularly affects the way we handle vehicular and pedestrian activity.

Vehicular activity

Usually, a major activity in the central parts of public spaces is vehicle circulation. Ways of designing so that vehicles do not inhibit other users of the space are discussed in Design Sheets 4.8 and 4.9.

Thun, Switzerland

Utrecht, Netherlands

Pedestrian activity

Most spaces are colonised from their edges. Where spaces are wide and there is no vehicular activity, the parts furthest from the edges may have little going on. Ways of animating the central parts of such places are explored in Design Sheet 4.10.

Colmar, France

The importance of microclimate

Finally, activities out of doors need appropriate microclimatic settings. These are covered in Design Sheet 4.11.

Place Dauphine, Paris,

... IN THE TOWN SQUARE, SUNLIGHT AND SHADE COULD BE FOUND AT THE SAME TIME + PLACE

Design implications

How to achieve robustness

1. Select the most robust configuration for any family houses in the scheme (Design Sheet 4.1)

2. In other buildings, locate all the elements of accommodation together in plan and section, working within the following constraints:
   - fit as much accommodation as possible into the preferred building configuration (Design Sheet 4.2)
   - make sure that ground floor areas abutting public outdoor space are occupied by active areas, so far as possible (Design Sheet 4.3)
   - position hard zones where they will not restrict the use of the remaining space (Design Sheet 4.4)

3. Adjust room sizes and details to maximise small-scale robustness (Design Sheet 4.5)

4. Design private open space for housing (Design Sheet 4.6)

5. Design the edges between buildings and public space to support as wide a range of likely uses as possible (Design Sheet 4.7)

6. Design all public spaces in detail, as follows:
   - busy vehicular streets (Design Sheet 4.8)
   - shared street spaces (Design Sheet 4.9)
   - pedestrian spaces (Design Sheet 4.10)

7. Check the microclimatic design (Design Sheet 4.11)

4.1: Robust family houses

In many areas, one of the commonest types of development is the family house with garden. This has a particular potential for robustness, which must be supported in the detailed design.

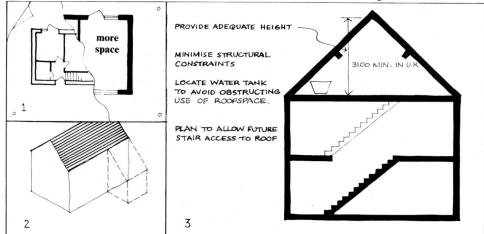

The most important factor affecting the robustness of a given house design is the area of space it provides (1). So robustness is supported by cheap straightforward construction, providing more space for a given cost; and by opportunities for enlarging the house as a whole (2). In all house types, this has implications for roof design: make sure its construction and geometry will allow easy conversion to useable space; and plan the original internal layout to make roof access easy (3).

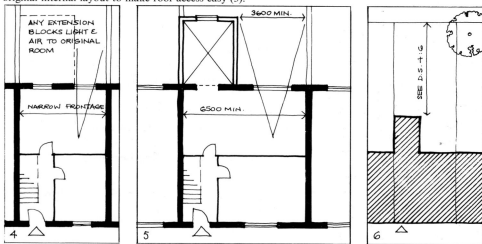

Terraced houses are the most difficult to extend horizontally, because additions can only be made at the front and back. With narrow frontages (4), below about 6.5 metres, front or back extensions begin to block light and air from the original rooms: easy roof conversion is *particularly* important here.
Extra frontage makes horizontal additions possible; but only if the original design allows for access (5). Garden sizes must allow for any potential horizontal extensions (6).

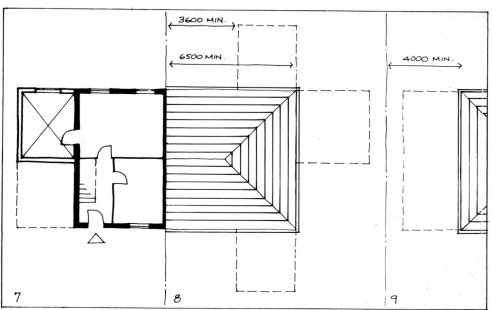

Semi-detached houses allow extensions at one side (7) and - with frontages above 6.5 metres - at front and back as well (8). A free space of at least 4 metres at the side is needed to allow pedestrian access past the minimum width of an average sized room extension (9).

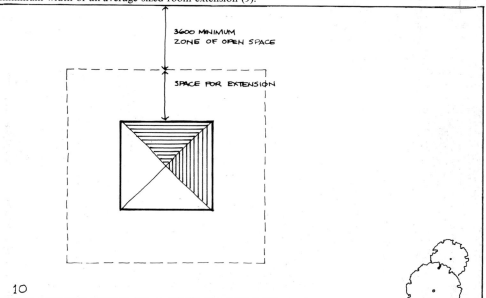

Detached houses offer more possibilities for extension than do other types. If the free space surrounding the house is at least 6.5 metres wide, an extension may be built onto any side (10). But remember that the first priority is to provide good space standards in the first instance, because not all users are able to extend their houses.

4.2: Preferred building configuration

The large-scale robustness of a building depends on three key factors:
- access
- depth from window to window
- height.

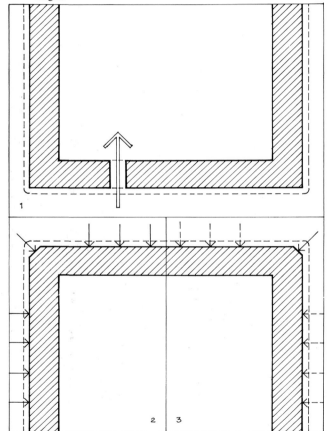

Make sure the building is sited to allow easy external access to as much as possible of the ground floor[6]. If there is no back access from the inside of the perimeter block, then provide a direct vehicular access to the back of the plot, from the public street, either past or through the building (1). If this is not required in the short term, at least make it possible later, without wholesale demolition. Maximise the direct frontage between building and public space, to allow as many separate front doors as possible. Again, if this is not necessary in the short term, make it easy to achieve later on (2,3).

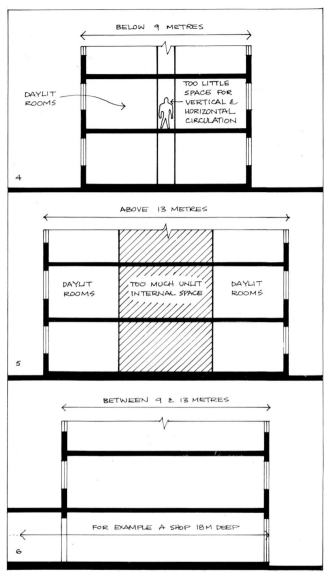

Indoor space is most robust when it can be naturally lit and ventilated. Robust buildings are therefore *shallow* in plan: the most robust depth is between 9 and 13 metres. Below 9 metres, the building is too shallow for a central corridor, and this limits the possible internal arrangements (4). Above 13 metres, the space is too deep to allow subdivision into small rooms, unless some are internal (5). So organise as much of the building as possible into a 9-13 metre depth, keeping those uses which will not fit as separate as possible. In this way, at least the major part of the building will have a high level of large-scale robustness (6).

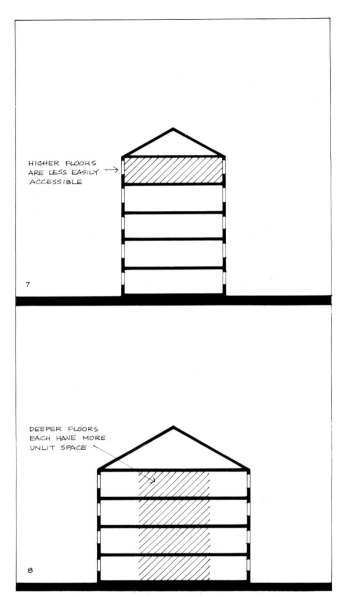

Next check how high this robust part of the building needs to be: robustness is reduced when the building is higher than 4 floors[7]. If there is too much accommodation to fit into this height, then two approaches are possible: make the building higher or increase its depth (7, 8). If the depth initially considered was less than 13 metres, then increase it to that figure. But if the area is still too small, it is better to increase height than to make the building still deeper. With a higher building, only the part above 4 floors loses robustness. With a deeper building, the robustness of *every* floor is reduced.

4.3: Active building fronts

The public edge of the building should house activities which benefit from interaction with the public realm, and can contribute to the life of the public space itself.

Courtenay Square, London, England

The first step is to locate as many entrances as possible in such positions that comings and goings are directly visible from public space (1).

South Molton St, London, England

The next step is to analyse the schedule of accommodation to see whether it contains any uses which could benefit from spilling out into public space (2). If such uses exist, then locate them on the ground floor at the front. If they require more space than is available in this location, still put the surplus at the front, but on the first floor (and if this is still too small, overflow to the second floor (3) and so on).

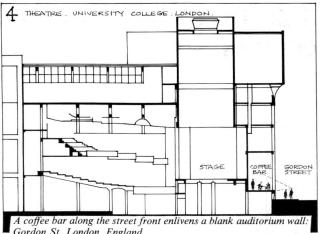

A coffee bar along the street front enlivens a blank auditorium wall: Gordon St, London, England.

Even if there are no uses which would benefit from links with public space, most buildings contain activities which can contribute to the animation of the public space itself (4,5). Make sure that these uses, rather than stores or lavatories, occupy the ground floor front position. Design Sheet 4.7 discusses how these activities can, if necessary, be protected from overlooking whilst still animating the public edge.

Oude Kerk, Amsterdam, Netherlands: exterior and plan.

6 a

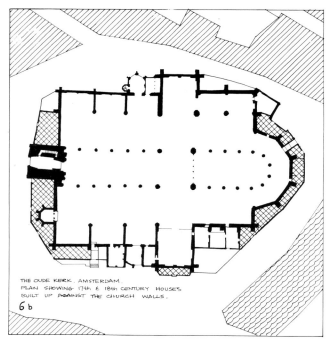

6 b

THE OUDE KERK, AMSTERDAM.
PLAN SHOWING 17th & 18th CENTURY HOUSES
BUILT UP AGAINST THE CHURCH WALLS.

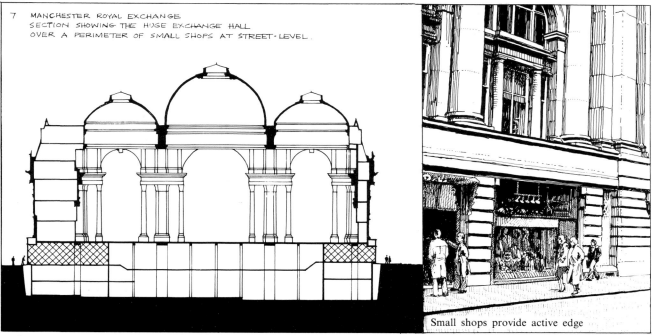

7 MANCHESTER ROYAL EXCHANGE
SECTION SHOWING THE HUGE EXCHANGE HALL
OVER A PERIMETER OF SMALL SHOPS AT STREET-LEVEL.

Small shops provide active edge

Single-aspect flats enliven the front of a multi-storey car park:
Birmingham, England.

8

If the building contains no uses which could contribute to animating the public edge, or if such uses as there are must be walled off for operational reasons, try to expand the schedule of accommodation beyond that originally envisaged by the patron, to include active edge uses (6,7,8). This will obviously require the patron's co-operation, which is more likely if the extra uses benefit the building itself in some way, financially or otherwise. Since the most 'blank' buildings - supermarkets, multi-storey carparks, theatres - are often in city centre locations, where space on the public front is particularly valuable, such benefit can be provided surprisingly often. Even if this can be achieved, there may still be occasional blank walls. Design Sheet 4.7 discusses how these may be turned to good account.

4.4: Interiors: large-scale robustness

Most buildings contain spaces which house shared facilities, such as staircases, lifts and major vertical service ducts. Usually these spaces are *hard:* they are the least likely to change their functions during the building's life. These hard zones must be positioned where they will not restrict the use of the remaining space.

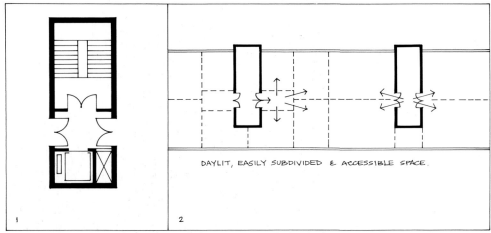

DAYLIT, EASILY SUBDIVIDED & ACCESSIBLE SPACE.

In buildings with less than about 15 metres frontage, group the hard zones together (1). Leave the rest of the space uninterrupted, so that it can easily be subdivided in many different ways (2), or split into separate units, such as flats or small office suites (3).

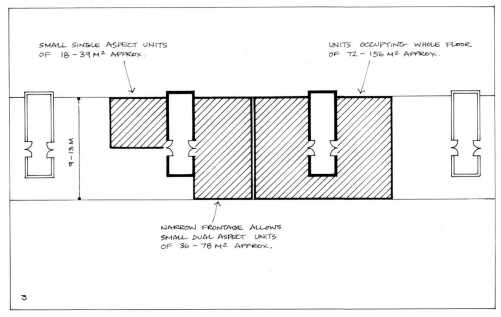

SMALL SINGLE ASPECT UNITS OF 18 – 39 M² APPROX.

UNITS OCCUPYING WHOLE FLOOR OF 72 – 156 M² APPROX.

9 – 13 M

NARROW FRONTAGE ALLOWS SMALL DUAL ASPECT UNITS OF 36 – 78 M² APPROX.

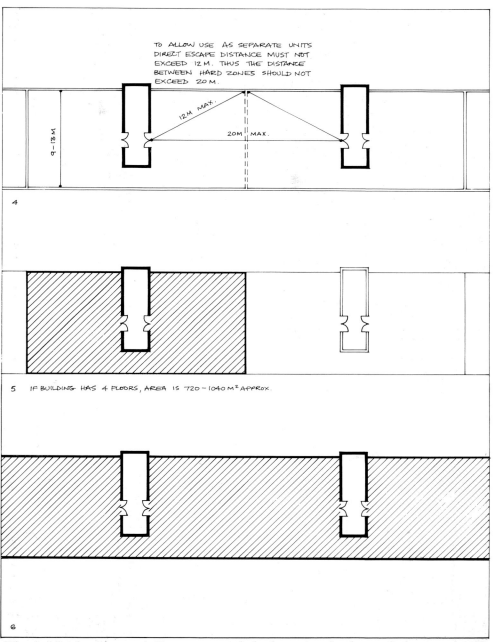

TO ALLOW USE AS SEPARATE UNITS DIRECT ESCAPE DISTANCE MUST NOT EXCEED 12 M. THUS THE DISTANCE BETWEEN HARD ZONES SHOULD NOT EXCEED 20 M.

9 – 13 M

12 M MAX.   20 M MAX.

4

5   IF BUILDING HAS 4 FLOORS, AREA IS 720 – 1040 M² APPROX.

6

When building frontage exceeds 15 metres, repeat the hard zones, spacing them no farther apart than 20 metres[8](4). Up to this spacing, the scheme can either be used as a number of *separate* buildings - each with adequate services, access and fire escape (5) - or can function equally well as a single unit (6). Whichever arrangement is required in the short term, it can be adapted later.

4.5: Interiors: small-scale robustness

Having developed the basic organisation of the building, the next step is to design individual spaces which, whilst being suitable for their initial purposes, are also capable of being put to the widest possible range of alternative uses.

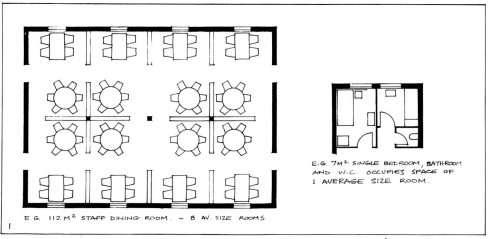

E.G. 7M² SINGLE BEDROOM, BATHROOM AND W.C. OCCUPIES SPACE OF 1 AVERAGE SIZE ROOM.

E.G. 112.M² STAFF DINING ROOM. - 8 AV. SIZE ROOMS

A very high proportion of the most common activities can fit into rooms about 14 m² in area: *average sized rooms* [9].

So use any freedom there may be in the schedule of room sizes in the brief, to make as many rooms as possible of this size (1). The dimensions of circulation spaces are also important for robustness: a small increase in the minimum circulation area may make them suitable for a wider range of activities, whilst still performing their basic linking functions (2,3).

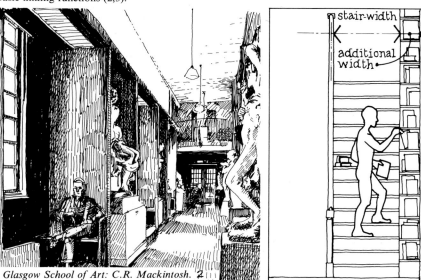

Glasgow School of Art: C.R. Mackintosh.

stair-width

additional width

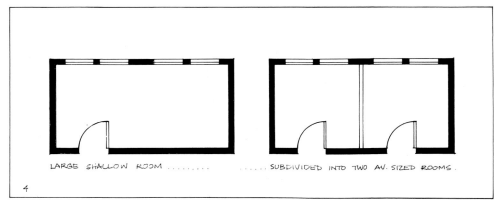

LARGE SHALLOW ROOM . . . . SUBDIVIDED INTO TWO AV. SIZED ROOMS.

4

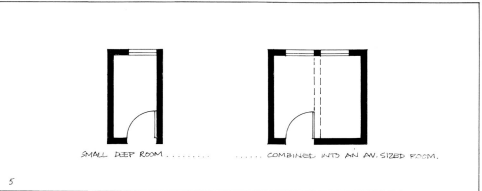

SMALL DEEP ROOM . . . . COMBINED INTO AN AV. SIZED ROOM.

5

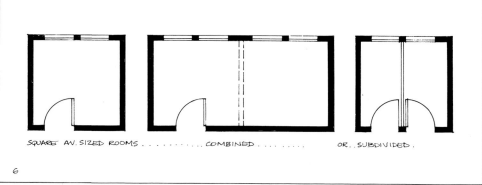

SQUARE AV. SIZED ROOMS . . . . COMBINED . . . .   OR. . SUBDIVIDED .

6

For a given area, rectangular rooms with plan proportions between 1:1 and 1:2 can accommodate the widest range of activities. *Shallow* rooms - with their windows on the long side - can easily be subdivided into spaces with natural light and useful proportions, whilst *deep* rooms are more easily combined into larger spaces of useful shape. So give any rooms which exceed 14 m² a shallow form, so that they can be subdivided into smaller ones (4). Make only the smaller rooms deep, and avoid separating them by structural walls, so they can easily be combined into average sized rooms if the opportunity arises (5). Average sized rooms themselves are best square, so they can be combined *or* subdivided (6).

Library, San Juan Capistrano, Ca., USA: Robert Stern. 7

A Lutyens hall: Little Thakeham, England. 8

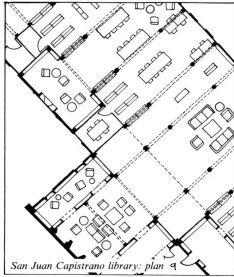

San Juan Capistrano library: plan 9

Larger spaces will seem appropriate to a wider range of uses if they can be read as being built up from a number of average sized spaces (7). Such spaces can provide an appropriately-scaled setting for individuals or small groups as well as being useful for larger gatherings (8). Rooms of this kind also lend themselves to physical subdivision, temporary or permanent, if required (9).

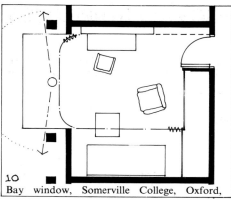

Bay window, Somerville College, Oxford, 10

A Voysey ingle: Chorleywood, England. 11

Mackintosh window seat: Helensburgh, Scotland. 12

Centraal Beheer, Appeldoorn, Netherlands: Herman Hertzberger. 13

The range of alternative settings within a given room or circulation space - and therefore its robustness - can also be increased by adding sub-spaces of different character, such as bay windows, ingles and window seats (10,11,12). These enable users to adopt a range of relationships to the main activity in the space: participation, observation, withdrawal. The sub-spaces can be quite small (13).

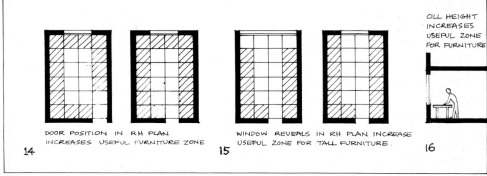

CILL HEIGHT INCREASES USEFUL ZONE FOR FURNITURE

DOOR POSITION IN RH PLAN INCREASES USEFUL FURNITURE ZONE 14

WINDOW REVEALS IN RH PLAN INCREASE USEFUL ZONE FOR TALL FURNITURE. 15

16

It is important that room design should maximise the opportunities for alternative furniture layouts. The most useful zone for furniture is round the edge of the room, so minimise the intrusion of doors into this area (14). For the same reason, provide reveals to windows (15,16) . Finally, remember that built-in furniture freezes room layout, and therefore reduces the range of possible uses. Avoid it where you can.

4.6: Housing: private gardens

Outdoor space which is private, within the perimeter block, greatly increases housing robustness. Detailed garden design should be left to individual users, but garden robustness is also affected by broader design issues.

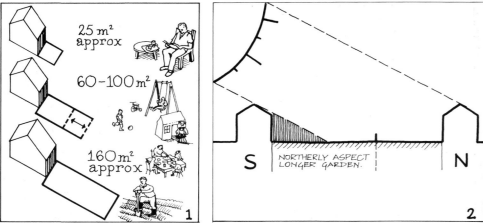

An area between 60-100 square metres is enough for both sitting out and for children's play, whilst 25 square metres is only adequate for passive activity. A family of four can become self-sufficient in vegetables with an area of 160 square metres[10] (1).

The best garden shape depends on the height and the aspect of the dwellings as these factors will affect sunlight. As a general rule, the more nearly the back of the dwelling faces north, the longer the garden must be to achieve adequate sunlight (2). But check each case with sunpath diagrams: see Design Sheet 4.11.

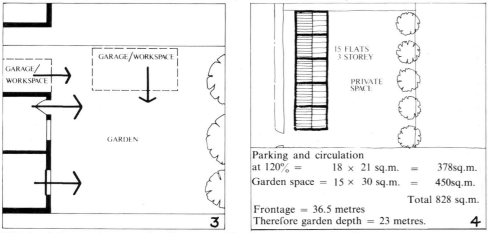

Parking and circulation
at 120% =   18 × 21 sq.m.   =   378sq.m.
Garden space = 15 × 30 sq.m.   =   450sq.m.

                                 Total 828 sq.m.
Frontage = 36.5 metres
Therefore garden depth = 23 metres.

With family houses, the garden should directly adjoin the back of the house, with easy access from as many rooms as possible (3). With flats the same applies for ground floor units, which can easily have private gardens: other flats can only have gardens separated from the dwelling. If at least 25 square metres is allowed for each upper-floor flat then the space can either be separate plots or a communal garden: users can have the choice. Add extra allowance for access paths, and for children's play (4).

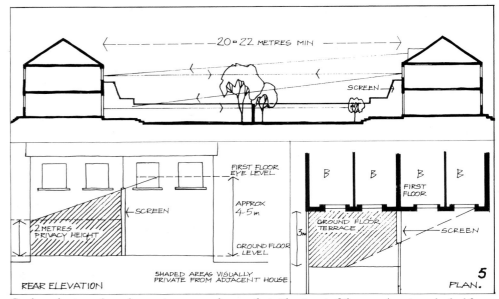

Garden robustness depends on *privacy,* so make sure that at least part of the space is not overlooked from next door, opposite or above. Always draw sections right across the block to check this, and to decide the positions and heights of fences, walls, pergolas, planting, canopies and garages to maintain privacy (5).

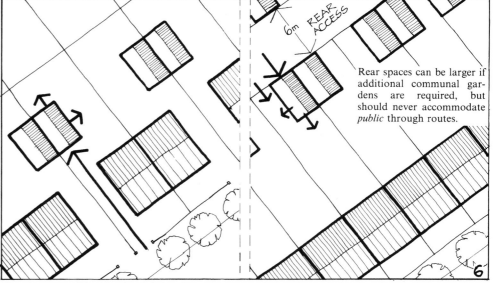

Rear spaces can be larger if additional communal gardens are required, but should never accommodate *public* through routes.

Rear access - or side access in semi-detached houses - increases the garden's potential for a range of activities, from horticulture to boatbuilding, because large and messy loads can be taken in and out without going through the house. If rear garages - or parking spaces which could later take garages - are directly attached to the garden, they can be used in conjunction with the garden itself for a wide range of extra activities. (6). But privacy - and therefore robustness - are destroyed if the back access becomes a public through route. So keep vehicle entrances small and obviously private.

4.7: The edge of the space

To increase robustness, the edge between buildings and public space must be designed to enable a range of indoor private activities to co-exist in close physical proximity with a range of outdoor public activities. This has a variety of design implications, depending both on the building activities concerned, and on the nature of the activities in the public space.

First, consider whether the building activity would itself benefit from claiming adjacent public space; and allow for it in the edge design. Common examples include residential balconies (1), terraces to pubs and restaurants (2), and display areas for shops (3).

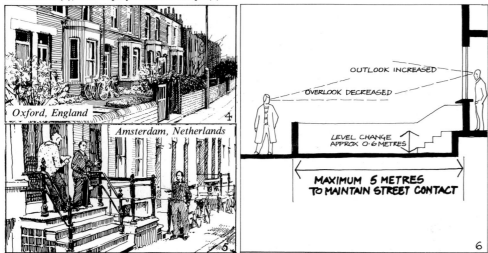

Elsewhere, however, an important function of the edge is to preserve the privacy of the indoor activity, so that users will not feel the need to screen themselves totally from the public space, thus negating any contribution that their presence might have made to the experience of the space itself. This privacy can be achieved by horizontal distance (4), level change (5), or a combination of both (6).

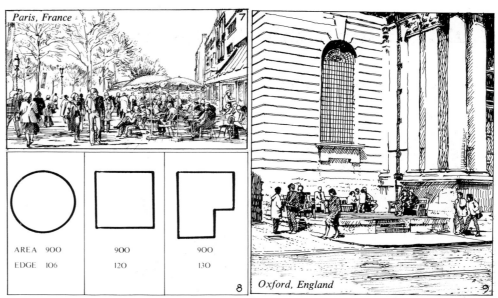

In locations where public activity is sufficiently intense, watching other people becomes in itself one of the most common activities. This mostly happens at the edge of the space, which offers a sense of refuge as well as a prospect of what is going on[11](7):the greater the proportion of edge to the area of the space, the greater the opportunities (8). The feeling of refuge can be increased by an indented building line (9). But be careful not to reduce the prospect by making the nook too deep.

The usefulness of the edge as a support for people-watching is greatly increased by the provision of places to sit. These need not always be single-purpose seats: if properly dimensioned, niches (10), string courses (11) and column bases (12) can work very well as seats and do not look forlorn when not in use.

City hall, Sheffield, England 13

Eguisheim, France 14

If seating is at a slightly higher level than the space itself then prospect is enhanced (13,14).

16

Veurne, Belgium

The edge potential is still further improved if parts of it can be protected from the weather (16).

Interlaken, Switzerland 15

This can be of commercial advantage to buildings which claim the edge (15).

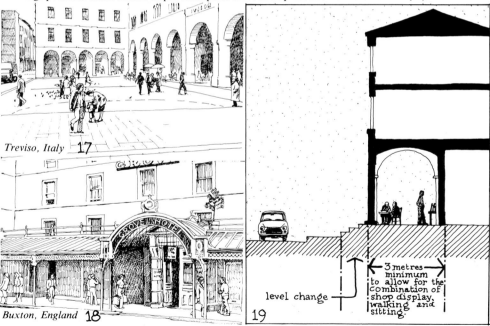

Treviso, Italy 17

Buxton, England 18

19

level change

3 metres minimum to allow for the combination of shop display, walking and sitting

Arcades are ideal for this (17, 18): the ultimate is the arcade raised above the level of the adjacent space to give a greater view (19).

4.8: Busy vehicular streets

Footpaths have a complex role to play in supporting pedestrian use against the inhibiting effects of vehicular traffic. In addition to the edge zone discussed in Design Sheet 4.7, they need two further zones: a central one for pedestrian movement, and a buffer zone between this and the vehicular space (1)[12].

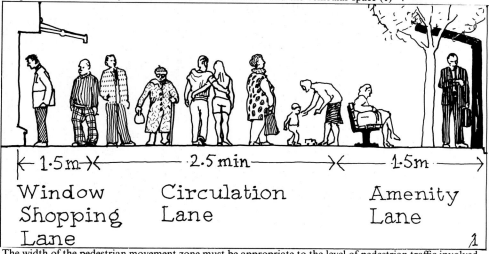

Window Shopping Lane | Circulation Lane | Amenity Lane

1

The width of the pedestrian movement zone must be appropriate to the level of pedestrian traffic involved. Between this movement zone and the vehicular space, allow a zone for amenities such as street trees, seating, bus shelters, telephone kiosks and cycle racks. Not all of these can be justified in all situations, but leave space for others to be added later.

Paris, France

2

Paris, France

3

Remember that parked cars are themselves one of the most effective barriers between pedestrians and *moving* vehicles (2). The types of roads in the scheme, as discussed in Design Sheet 1.3, will determine whether on-street parking is possible. If not, it may be worth making an extra parking slip road (3). But make sure that this will not contradict the decisions about street enclosure made in Design Sheet 3.6.

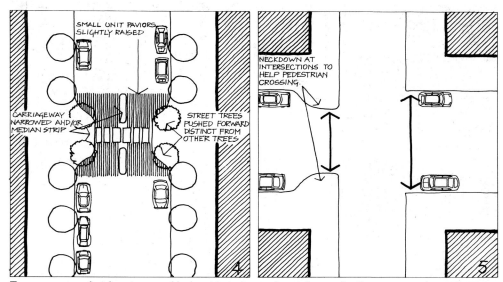

4

5

To encourage pedestrians to use vehicular streets, it must be made easy for them to cross the road. Most people prefer to cross at ground level, rather than by subways or bridges. Provide safe crossing points (4), making them as visually prominent as possible (5), and minimise road widths at these positions. All crossing points should cater for handicapped people, and where there are traffic signals these should be timed in favour of the pedestrian as far as possible.

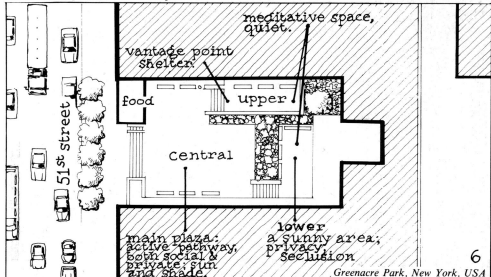

6

Greenacre Park, New York, USA

In noisy vehicular streets, seclusion and quiet can be achieved by providing relatively small spaces, set back from the building line. To benefit as many people as possible, these spaces are best located in areas of high pedestrian activity. To reduce danger and vandalism at night, they should be brightly lit; and either surrounded by buildings with night-time as well as day-time use, or be tightly managed in security terms. To reduce the impact of traffic, the open frontage of such a space is best kept to the minimum necessary to maintain a street prospect. This is improved if the space is raised above street level (6)

4.9: Shared street spaces

In some situations - mostly residential - and with careful detailed design, streets can be made robust enough for the space to be *shared* by vehicles and pedestrians.

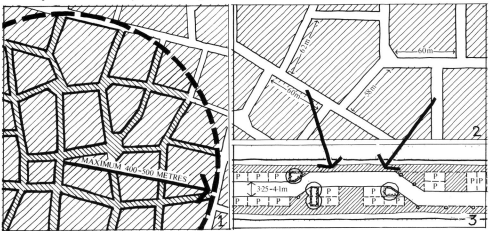

The shared street is only possible where traffic flows are less than 250 vehicles per hour, and the majority of the traffic has its destination within the area itself. No area of streets designed on the 'shared space' principle should be more than 500 metres from a 'normal' vehicular street (1). Each street in the area should have directional changes every 50 to 60 metres: the small-block structure advocated in Chapter 1 is ideal (2), but additional changes in direction may be necessary (3). Two-way traffic should be encouraged throughout the area, to reduce vehicle speeds.

Utrecht, Netherlands

The road section should be kept narrow (4), with occasional widening for passing places, rather than the other way round. Adequate parking for residents and visitors must be provided, and on-street parking should be of right-angle form: this demands greater attention from drivers, and provides better play spaces when cars are absent (5).

Delft, Netherlands

Kerb distinctions, with their emphasis on *separating* vehicles from pedestrians, should be eliminated, and replaced by paving to reduce the linearity of the space. There must be many elements to reduce vehicle speeds, but, rather than arbitrary devices, it is important that these are seen by drivers as giving *advantages* to other users of the area: trees, children's play equipment or parked cars (6). But it is important that drivers can see children: raised obstructions should be lower than 750 mm (7).

Utrecht, Netherlands

The design ideas outlined above are all used in the 'Woonerf' streets now common in the Netherlands . They may sometimes conflict with traffic regulations elsewhere, but they illustrate what is functionally practicable when regulations can be relaxed (8)[13].

4.10: Pedestrian spaces

Only exclude vehicles from a public space in the following situations:
- if the vehicles inhibit pedestrian activity (which is rarely the case, except in busy commercial streets)
- if there is an alternative vehicular route near by.

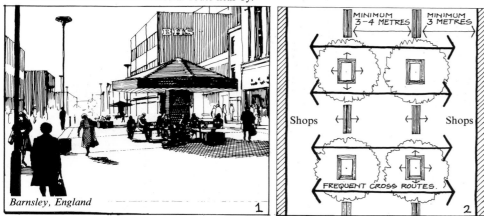

Barnsley, England 1

Where pedestrian streets are less than 7 metres wide, design their edges as described in Design Sheet 4.7. Wider streets or pedestrian squares need further supports for people to colonise the centre of the space. These should incorporate seating (1). Locate seating parallel to pedestrian flows: on wider streets with active uses on both sides, arrange seating down the centre of the space (2). In squares establish likely desire lines for pedestrian flows and then arrange seating to take advantage of the people-watching potential of these positions. Remember that people also like to stand or lean in similar locations (3).

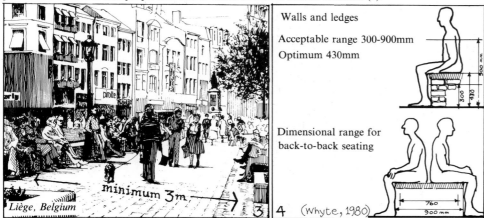

Liège, Belgium 3   4   (Whyte, 1980)

Seating can take the form of chairs or benches (often called primary seating) or secondary features like steps, walls or planters. Provide as much primary seating as possible - never less than 10% of the total number of seats - and ensure that there is at least 300 linear mm of seating for each 3 square metres of open space(4). It is important to allow for choices of seating configuration, with as much seating as possible raised a little above the level of the surrounding paved areas. Avoid locating seats *lower* than their surroundings, as this markedly reduces their potential prospect. Include moveable seats and tables wherever management can cope. The implications of various seating arrangements are explored in diagram 5.

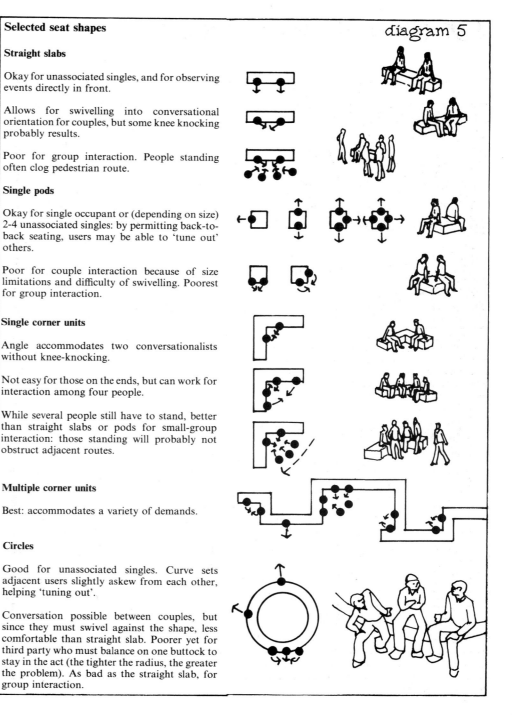

Selected seat shapes

Straight slabs

Okay for unassociated singles, and for observing events directly in front.

Allows for swivelling into conversational orientation for couples, but some knee knocking probably results.

Poor for group interaction. People standing often clog pedestrian route.

Single pods

Okay for single occupant or (depending on size) 2-4 unassociated singles: by permitting back-to-back seating, users may be able to 'tune out' others.

Poor for couple interaction because of size limitations and difficulty of swivelling. Poorest for group interaction.

Single corner units

Angle accommodates two conversationalists without knee-knocking.

Not easy for those on the ends, but can work for interaction among four people.

While several people still have to stand, better than straight slabs or pods for small-group interaction: those standing will probably not obstruct adjacent routes.

Multiple corner units

Best: accommodates a variety of demands.

Circles

Good for unassociated singles. Curve sets adjacent users slightly askew from each other, helping 'tuning out'.

Conversation possible between couples, but since they must swivel against the shape, less comfortable than straight slab. Poorer yet for third party who must balance on one buttock to stay in the act (the tighter the radius, the greater the problem). As bad as the straight slab, for group interaction.

diagram 5

Selected seating arrangements

Strict linearity

Distance from events for detached viewing, to enable seat to form a 'refuge'.

People on ends can swivel easily for conversation.

Person on end can turn back on immediate neighbours, to 'tune them out', without making eye contact with people on next bench.

1.2m. maximum, to allow interaction between users.

Where rows of benches flank a passageway, place them at least 3m. apart, so that interaction between people sitting on opposite rows will not make the passageway awkward for others to use.

Right angles

Avoid clumsy overlaps.
Similar distances apply as with strict linearity.

Clusters

Vary as much as possible to accommodate many combinations of distance and orientation, including service for the occasional loner.

Source: Rutledge, 1980

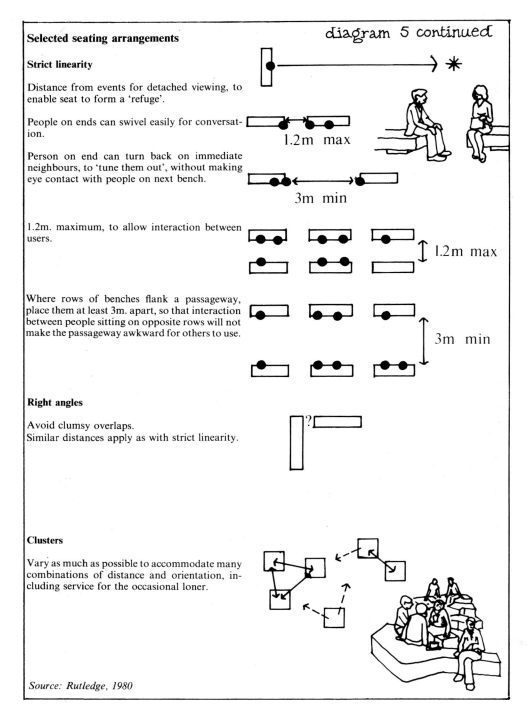

diagram 5 continued

1.2m max

3m min

1.2m max

3m min

?

Liège, Belgium. 6

Market as a series of small double-loaded streets, creating a high proportion of new *edges* within the space. 7

min 3m

3-4m

Sheffield, England 8

Seating is not the only means of encouraging people to colonise the centres of spaces. A pavilion building of public use - or even a monument (6) - could provide the necessary supports. At a smaller scale, market stalls and information kiosks can draw people to the centre, particularly if they incorporate seating and shelter (7,8).

2.5 m. minimum   *Montmartre, Paris, France* 9

Begin with this as a principle...

10 ...and alter as other requirements demand

11

Trees can form smaller enclosures within the main space: like edges, these can provide the combination of refuge and prospect which encourages people to claim the space. The base of the tree canopy must be at least 2.5 metres above ground level (9). To make a robust series of spaces between trees, plant them on a roughly square grid about 5 metres apart. This will form the outdoor equivalent of the 'average sized rooms' discussed in Design Sheet 4.5, each capable of supporting a wide range of activities without obstructing pedestrian movement. Do not worry that a grid pattern will seem monotonous: it is simple in plan, but complex in perspective (10,11).

4.11: Microclimate

The range of activities in an outdoor place - and hence its robustness - depends partly on its microclimate: particularly windspeed and sunlight. Begin by asking the local meteorological office for data on windspeed and direction, for the weather station nearest to the site.

Situation	Windspeed m/s	Effect
Calm, light air	0-1.5	Calm, no noticeable wind
Light breeze	1.6-3.3	Wind felt on face
Gentle breeze	3.4-5.4	Wind extends light flag, hair is disturbed, clothing flaps.
Moderate breeze	5.5-7.9	Raises dust, dry soil, loose paper; hair disarranged.
Fresh breeze	8.0-10.7	Force of wind felt on body, drifting snow becomes airborne, limit of agreeable wind on land.
Strong breeze	10.8-13.8	Umbrellas used with difficulty, hair blown straight, difficult to walk steadily, wind noise on ears unpleasant, windborne snow above head height (blizzard).
Near gale	13.9-17.1	Inconvenience felt when walking.
Gale	17.2-20.7	Generally impedes progress, great difficulty with balance in gusts.
Strong gale	20.8-24.4	People blown over by gusts

After Penwarden and Wise, 1975   **1**

Windspeed is important partly because it affects temperature. For example, a 50 kph wind at minus one degree centigrade has six times the cooling effect of still air at minus twelve degrees centigrade. Fig. 1 summarises the effects of windspeed on human activity, and therefore on the robustness of the space. The clear implication is that we should design to keep windspeeds below 5 metres per second[14].

Figure 2 indicates typical effects of building form on windspeed. Note the considerable advantages of four-storey buildings in street/block configurations, as advocated in Chapters 1-4.

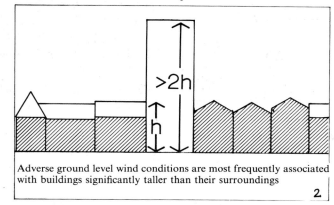

Adverse ground level wind conditions are most frequently associated with buildings significantly taller than their surroundings

2

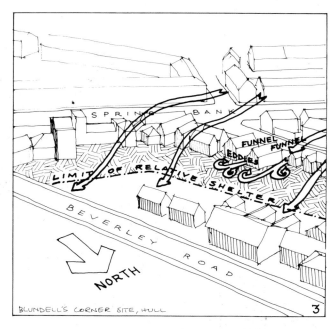

BLUNDELL'S CORNER SITE, HULL   **3**

To be absolutely sure that windspeed problems are minimised, use a wind-tunnel to experiment with possible improvements. This will require a model of the site and its surroundings, covering a scale radius of at least 100 metres, at a scale of at least 1:200. Figure 3 shows the type of information which such a test can contribute.

People tend to follow the sun across a space; seeking or avoiding it according to the climate. The amount and position of sunlight in the space depends on the latitude concerned (4a,b,c).

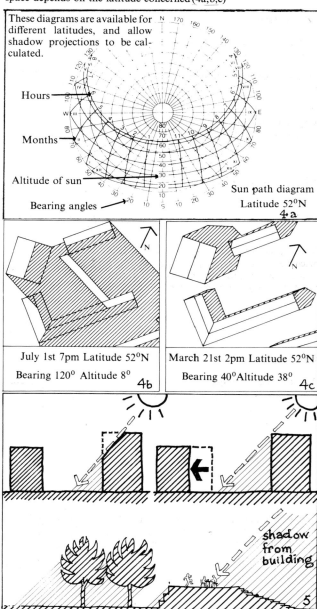

These diagrams are available for different latitudes, and allow shadow projections to be calculated.

Hours
Months
Altitude of sun
Bearing angles

Sun path diagram
Latitude 52°N

4a

July 1st 7pm Latitude 52°N
Bearing 120° Altitude 8°

4b

March 21st 2pm Latitude 52°N
Bearing 40° Altitude 38°

4c

shadow from building

5

The areas of sunlight and shade can be altered by design adjustments at a variety of scales: building mass, open space width, level changes, trees or other features within the space (5).

Chapter 5: Visual appropriateness

Introduction

Our earlier decisions about planning and massing have determined what the design should look like in general terms. Now we must focus on its appearance in more detail.

This is important because it strongly affects the *interpretations* people put on the place: whether designers want them to or not, people *will* interpret places as having meanings[1]. When these meanings support responsiveness, the place has a quality we call *visual appropriateness*.

When is this important?

Visual appropriateness is particularly important in the places which are most likely to be frequented by people from a wide variety of different backgounds; particularly when the place's appearance cannot be altered by the users themselves. Both indoors and out, therefore, visual appropriateness is mostly important in the more *public* spaces of the scheme. So far as public outdoor space is concerned, it is particularly relevant to the outside of the buildings which define the public realm.

What makes the visuals appropriate?

The interpretations people give to a place can reinforce its responsiveness at three different levels:
- by supporting its legibility, in terms of form and use.
- by supporting its variety.
- by supporting its robustness, at both large and small scales.

Legibility of form

In Chapter 3, we designed the *mass* of the building to reinforce the legibility of the area in which it is located. The detailed appearance must now be designed to reinforce this objective. For example, if the building is intended to be visually integrated into its surroundings, it is important that users should interpret its detailed design as having a family resemblance to the buildings around it.

But there is a problem here: different groups of users may have different opinions about whether two given buildings share a similar character or not. One group may pay a great deal of attention to proportions, and to overall visual structure; whilst another may depend on more detailed cues: similarity in window and door design, for example.

Legibility of use

In Chapter 3, we considered how to locate uses to improve their legibility. The detailed appearance of the place must help people read the pattern of uses it contains. For example, a town hall should *look* like a town hall, and a house should *look* like a house, to as many users as possible. But here too there is a problem: what looks like a town hall to one group of people may look like a factory to others.

Or people may read it as a town hall, but still interpret it as an inappropriate *kind* of town hall: bureaucratic rather than democratic, for example. If a place is interpreted in this negative way, its users are far less inclined to adopt an active and exploratory attitude towards it. Its potential for responsiveness is correspondingly reduced.

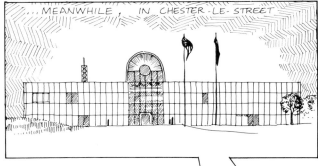

Variety

In Chapter 2, we considered how to make it possible for a wide variety of uses to co-exist in an area. The detailed appearance of the buildings must help this to happen, by making the image of the area seem appropriate as a setting for each of the uses concerned.

There is a problem here as well. People may not mind having a town hall across the street, but if they interpret it as a factory, they may be less enthusiastic.

Large-scale robustness

In Chapter 4 we considered how a building could be designed to accommodate a wide range of uses. Its detailed appearance must reinforce this potential, by looking appropriate for all these uses.

Here there is yet another problem: how can a building be designed to look like several things at once? And how can it still make clear what it *actually* houses at any given time?

Couldn't someone start a vogue, asks Eric de Maré, for transforming these riverside **Lincolnshire warehouses** *into modest latter-day northern versions of those Venetian palazzi?*

Small-scale robustness

At a smaller scale, Chapter 4 also considered ways of designing particular spaces within a building, or out of doors, so they could be used in a range of different ways. Thus a given house, for example, could be used by people with a wide range of different lifestyles. But this too raises problems: how can such a building be designed so that people from a range of different backgrounds will each see it as an appropriate home?

The bay windows make it look really homely

I like it ... it's most imposing

The role of detailed appearance

It should by now be clear that by overcoming these problems the detailed appearance of the scheme has an important role to play in supporting responsiveness: it is neither a mere by-product of the plan, nor a matter of artistic whim.

This idea that elevations have specific tasks to perform is an unfamiliar concept to most designers. To stop it being forgotten, it is necessary to write a detailed performance specification for the objectives which each of the scheme's publicly-visible surfaces is to achieve. This is covered in Design Sheet 5.1.

How can these objectives be achieved?

To encourage these interpretations to be made, we must understand how people interpret places.

How do people interpret places?

People interpret visual cues as having particular meanings because they have *learned* to do so[2]. But people do not learn in a social vacuum. A great deal of learning, both formal and informal, is shared by groups of people; whose members will therefore tend to make similar interpretations of a given place[3].

But members of different social groups may well make different interpretations of the same place. This happens for two main reasons:
- their environmental *experience* differs from that of other groups[4].
- their *objectives* differ from those of other groups[5].

For example, many British people have been brought up in streets like the one sketched below.

This is the predominant kind of housing which such people have experienced; so new buildings which contain similar visual cues, like the one sketched below, will probably be interpreted as housing.

Housing Project,
Swindon, England

But it is people's objectives which determine whether a building which looks like this will be interpreted as *appropriate* housing or not. For example, some groups may be intensely concerned with changing their social status, and may regard housing which takes its cues from the traditional street as inappropriately working class. For others, it may have a comforting familiarity.

This means that if we are to design visually appropriate places, using cues which different groups of users are likely to interpret as supporting legibility, variety and robustness, we have to enquire into the likely experience and objectives of the place's users, looking for visual cues relevant to each user group.

Which cues do we need?

To support legibility, we need cues which will be interpreted as relating the building concerned to its context: either reinforcing or standing out from the paths, nodes, landmarks, edges or districts concerned. We call these *contextual cues*. Variety and robustness, on the other hand, are both concerned with the ways in which the project is *used*. To support these qualities, we need cues which will be interpreted as appropriate to the various uses concerned. We call these *use cues*. Ways of finding appropriate cues of both kinds are discussed in Design Sheet 5.2.

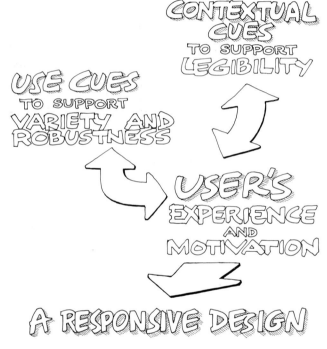

Using the cues in design

Once these sets of cues have been found, the final step is to use them to attain the objectives outlined in Design Sheet 5.1; employing the contextual cues to achieve objectives about legibility, and the use cues to support objectives about variety and robustness.

Ways of using contextual cues are explored in Design Sheets 5.3 and 5.4, whilst Sheet 5.5 considers use cues. The final process of bringing all the cues together into a detailed design is covered in Design Sheet 5.6.

Design implications

How to achieve visual appropriateness

1. **Take the design from Chapter 4 as the starting point for developing visual appropriateness.**

2. **Establish detailed objectives for each publicly-visible surface in the scheme, specifying which of the responsive qualities are to be communicated to each relevant user group (Design sheet 5.1)**

3. **Find the necessary vocabulary of contextual cues and use cues needed to achieve these objectives (Design sheet 5.2)**

4. **Consider implications of contextual cues in achieving objectives about legibility (Design Sheets 5.3, 5.4)**

5. **Consider implications of use cues in achieving objectives about variety and robustness (Design Sheet 5.5)**

6. **Employ contextual cues and use cues in the final design of each surface (Design Sheet 5.6)**

5.1: Detailed appearance: a specification

We have already designed the location and massing of the various parts of the project to support responsiveness. The next step is to design its publicly-visible surfaces to communicate the project's variety, legibility and robustness to a wide range of users.

Begin by making a drawing to show all the surfaces - elevations, roofs and floorscapes - whose detailed design is to be considered at this stage. Axonometrics are quick and useful for showing all these surfaces on one drawing.

Next consider all the publicly-visible surfaces, one by one, to decide which qualities each surface should communicate to whom. For each surface work through the qualities one at a time, as outlined below; remembering that every quality is not necessarily relevant to every surface.

Variety

- consider all the surfaces in the scheme which might need special design attention to prevent the image of one proposed use being seen as an inappropriate setting by its neighbours.
- for each of these surfaces, define the development agencies which might be put off by an inappropriate setting.
- record the design objectives implied by these factors, as illustrated on the drawing below.

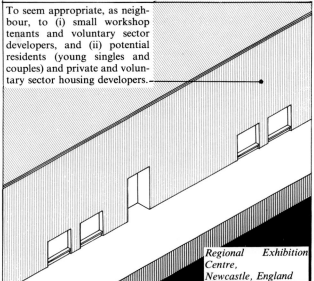

To seem appropriate, as neighbour, to (i) small workshop tenants and voluntary sector developers, and (ii) potential residents (young singles and couples) and private and voluntary sector housing developers.

Regional Exhibition Centre, Newcastle, England

Legibility

Begin by recalling the decisions about legibility of form made in Design Sheets 3.5, 3.6, 3.7 and 3.8. Then consider each publicly-visible surface in the scheme, recording any of these decisions which are relevant to it, as illustrated in the example below.

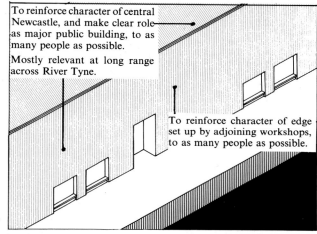

To reinforce character of central Newcastle, and make clear role as major public building, to as many people as possible.

Mostly relevant at long range across River Tyne.

To reinforce character of edge set up by adjoining workshops, to as many people as possible.

Small-scale robustness

- review the proposed uses of the indoor and outdoor spaces defined by each surface.
- consider all the interest groups using the building or outdoor space concerned; paying particular attention to those who might be dissuaded by an inappropriate image.
- record the design objectives implied by these factors, as illustrated on the drawing below.

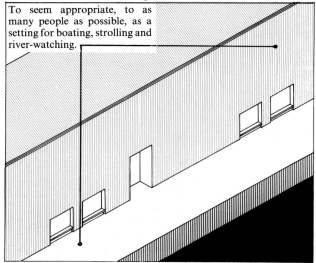

To seem appropriate, to as many people as possible, as a setting for boating, strolling and river-watching.

Large-scale robustness

- review the likely future uses -as explored in Chapter 4 - for the interior and exterior spaces defined by each publicly-visible surface.
- consider the interest groups who would benefit from being aware of the place's potential for these uses (this will include developers, tenants and purchasers of space for the uses concerned).
- record the design objectives implied by these factors, as illustrated in the drawing below.

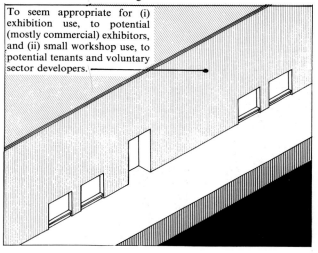

To seem appropriate for (i) exhibition use, to potential (mostly commercial) exhibitors, and (ii) small workshop use, to potential tenants and voluntary sector developers.

The full specification

Roof:
To reinforce character of central Newcastle, and make clear role as major public building, to as many people as possible.

Wall:
To seem appropriate, as neighbour, to (i) small workshop tenants and voluntary developers, and (ii) potential residents (young single persons and couples) and private and voluntary sector housing developers. To reinforce visual character of edge set up by adjoining workshops, to as many people as possible. To seem appropriate, to as many people as possible, as a setting for boating, strolling and river-watching. To seem appropriate for (i) exhibition use, to potential (mostly commercial) exhibitors, and (ii) small workshop use, to potential tenants and voluntary sector developers.

Floorscape:
To seem appropriate, to as many people as possible, as a setting for boating, strolling and river-watching.

By this stage each publicly-visible surface will have its set of objectives. The next step is to look for visual cues with which to achieve these objectives, as discussed in Design Sheet 5.2.

5.2: Looking for visual cues

Design Sheet 5.1 worked out how we want the scheme to be interpreted by various interest groups. The next step is to look for cues to support these interpretations. These cues are of two kinds:
- cues associated with particular *places*, which are needed to achieve objectives about legibility. We call these *contextual cues*.
- cues associated with particular *uses*, which are needed to achieve objectives about variety and robustness. We call these *use cues*.

Contextual cues

In Design Sheet 5.1, we considered each publicly-visible surface in the scheme, and decided whether to associate it with any particular path, node, landmark, edge or district. Now we must look for visual cues associated with any such context.

Begin by deciding where to look for these cues. Where the objective is to make the surface reinforce - or stand out from - a particular path or edge (as decided in Design Sheets 3.3 and 3.6) look for cues in that section of the path or edge from which the relevant surface can be *seen*, as sketched below.

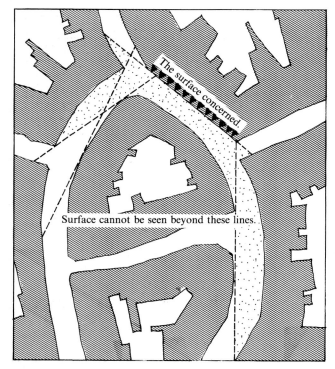

Surface cannot be seen beyond these lines.

Where the objective is for the surface to act as - or reinforce - a landmark (as decided in Design Sheets 3.3 and 3.8) look for cues in the area from which it will be seen, as sketched below.

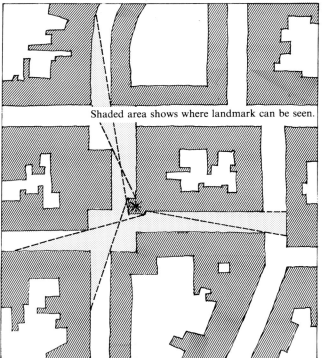

Shaded area shows where landmark can be seen.

Where the objective is to make the surface reinforce - or stand out from - some particular node or district, as decided in Design Sheets 3.4, 3.5 and 3.7, look in that area itself for cues to be *used*, and in adjoining areas, with which it might be confused, for cues to be *avoided*.

In each case, draw the relevant boundaries on a plan. Then walk round the relevant areas, looking for cues: any recurring visual features which people are likely to notice. The list below, compiled from various studies from a variety of cultures and backgrounds[6], forms a useful checklist of noticeable features.

- vertical rhythms
- horizontal rhythms
- skylines
- wall details (material, colour, patterning etc.)
- windows
- doors
- ground level details

Work through all the factors in the list, noting in sketches or photographs any recurring forms which exist. It is important to record these cues in an organised way, so that they can easily be referred to when designing. A useful format is illustrated on the next page: it shows some of the cues to be used, and others to be avoided, to reinforce the character of a particular district.

It is difficult to predict which *particular* interest groups will notice which *particular* cues. To include cues relevant to a wide range of groups, therefore, aim for a design vocabulary which includes as many as possible of the different types of features listed above. Finally, consider whether any of the cues you have found might be interpreted as *negative* by any of the interest groups thought relevant to the particular objective concerned. If so, avoid these cues when designing. Where resources permit, discuss such matters with representatives of the relevant interest groups.

Use cues

In Design Sheet 5.1, we worked out objectives about variety and robustness. Now we must find use cues to support these.

Most uses are associated with a variety of building images, each with its associated range of visual cues. For each of the objectives decided in Design Sheet 5.1, check through the list of relevant interest groups, to decide which of these building images is likely to be *familiar* to each group, and which they might *aspire* to: these are probably not the same. Again, check your own ideas about this with representatives of the interest groups concerned, so far as resources permit.

For each interest group, find as many examples of both the *familiar* and the *preferred* images of the relevant use as you have time for. Then use the checklist of noticeable features, as explained above, to generate a set of potential cues which could be used in design: the process is similar to that illustrated on the next page.

Housing images, Newcastle, England

By this stage, we have focussed on the cues which are associated with the particular contexts and uses which are relevant to the objectives decided in Design Sheet 5.1. Various ways in which these cues can be used in different situations are discussed in Design Sheets 5.3, 5.4, 5.5 and 5.6.

	Large scale cues				Small scale cues	
	Typical of own district	**Typical of adjoining districts avoid where possible**			**Typical of own district**	**Typical of adjoining districts avoid where possible**
Vertical rhythms				**Windows**		
Horizontal rhythms				**Wall details**		
Skylines				**Ground level details**		

5.3: Contextual cues: the surrounding area

In Design Sheet 5.1, we decided whether the new surface should reinforce or contrast with the visual character of its context; which was itself investigated in Design Sheet 5.2. Now we shall explore ways of using cues to achieve either objective.

The cues which were found when analysing the visual character of the context, in Design Sheet 5.2, were of two kinds:
- elements (such as wall details, windows, and door and ground-level details).
- relationships between elements (such as vertical or horizontal rhythms, and skyline relationships).

Both elements and relationships can vary from being all similar to being all different. It is useful to consider the four key possibilities illustrated below:

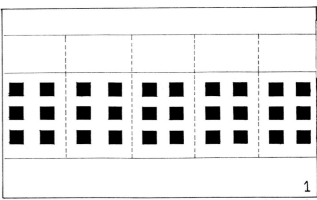

A visual character formed by *similar elements* arranged in *similar relationships* is illustrated above(1).

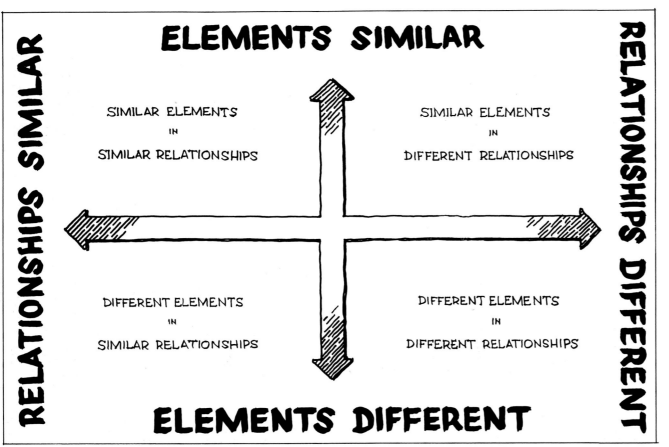

ELEMENTS SIMILAR

RELATIONSHIPS SIMILAR

SIMILAR ELEMENTS
IN
SIMILAR RELATIONSHIPS

SIMILAR ELEMENTS
IN
DIFFERENT RELATIONSHIPS

DIFFERENT ELEMENTS
IN
SIMILAR RELATIONSHIPS

DIFFERENT ELEMENTS
IN
DIFFERENT RELATIONSHIPS

RELATIONSHIPS DIFFERENT

ELEMENTS DIFFERENT

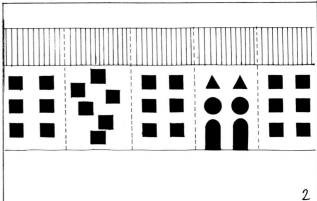

In this situation, the introduction of new relationships and/or new elements will make the new surface stand out from its context (2).

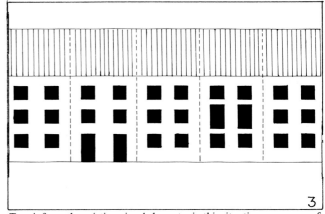

To reinforce the existing visual character in this situation, use many of the existing elements and relationships in the new design (3).

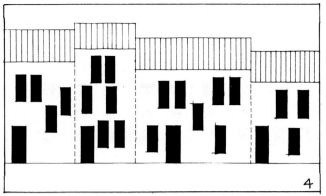

A visual character formed by *similar elements* arranged in *different relationships* is illustrated above (4).

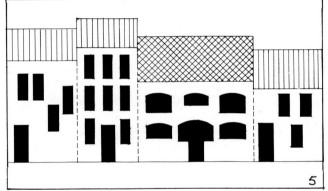

In this situation, contrasting *elements* will have more effect than contrasting *relationships*, in making the new design stand out from its context (5).

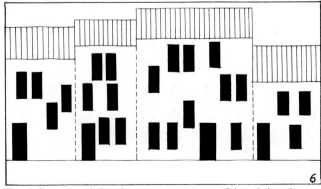

To reinforce the existing character, use as many of the existing *element* cues as possible in the new design (6). But the relationships between them need only be tentatively decided: they can be adjusted later to make the design richer, as discussed in Chapter 6.

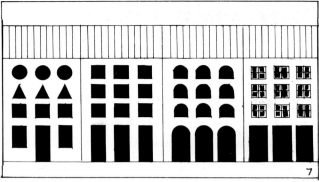

When there are many *relationships* as cues, but few common *elements*, the visual character is formed by different elements arranged in similar relationships (7).

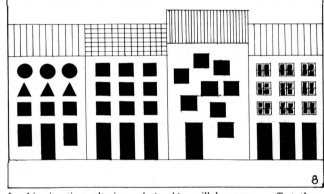

In this situation, altering *relationships* will have more effect than altering *elements*, in making the new design stand out from its context (8).

To reinforce the existing character, use as many *relationship* cues as possible(9).But the *elements* need only be tentatively decided: they can be adjusted to make a richer design, in Chapter 6.

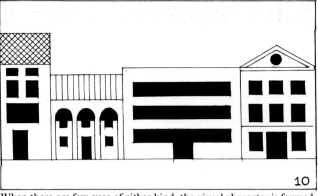

When there are few cues of either kind, the visual character is formed by different elements arranged in different relationships (10).

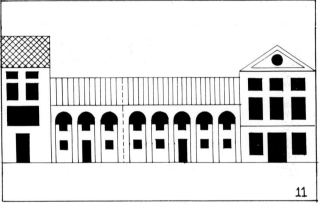

In this situation, a new design with either recurring elements or recurring relationships will contrast with the existing context (11).

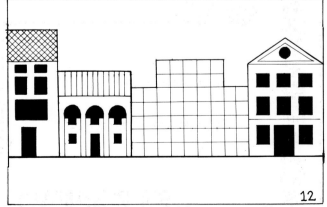

To reinforce the existing character, it is important to avoid setting up similar relationships or similar elements within the new design (12)

5.4: Contextual cues: the adjacent buildings

This Design Sheet suggests how to use a new design to unite adjoining buildings of disparate character, when this has been decided as an objective in Design Sheet 5.3. The important cues are those from the adjacent buildings: it is they which have the most direct visual relationship with the new design.

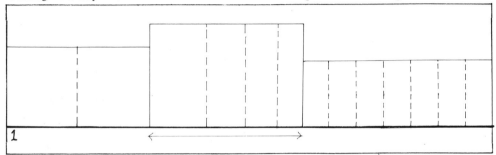

Start with large-scale cues. If the buildings on either side have any such cues in common, use these as a starting point (1). If they do not, see if you can use some from one side, and some from the other (2).

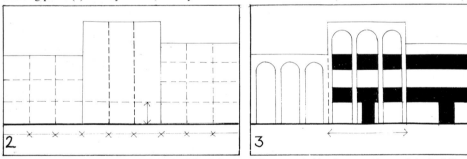

A further approach is to use the new building's large-scale cues to bridge between those of the buildings on either side (3).

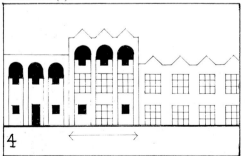
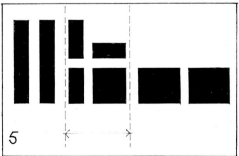

Next consider the smaller-scale cues. There are two possibilities here:
- Use cues from both sides, but particularly from the side whose large-scale cues were least used (4).
- make gradual transitions between cues on either side (5).

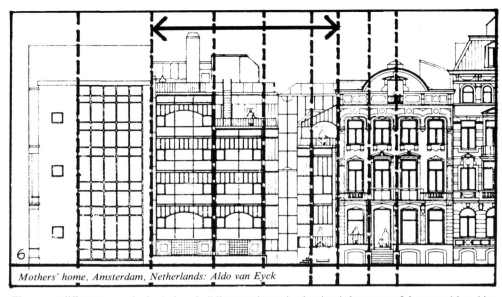

Mothers' home, Amsterdam, Netherlands: Aldo van Eyck

These very different examples both show buildings used to unite the visual characters of those on either side (6,7).

Bruges, Belgium

5.5: Use cues: supporting variety and robustness

This Design Sheet suggests ways of combining the use cues from Sheet 5.2 to achieve the objectives about variety and robustness developed in Sheet 5.1.

- Begin by analysing any large-scale patterns of buildings in the sets of cues for each use - skylines, vertical and horizontal relationships and so forth - looking for cues in one set which are similar to those in the others, and which could therefore be used to form the main visual structure of the design. This ability to recognise useful similarities is more of an art than a science: it develops with practice.
- Once the large-scale skeleton of the design has been established, work down through the smaller-scale cues - for example windows, entrances and ground-level details - again looking for similarities among cues for the various uses, to develop a detailed design for the surface concerned.
- Finally, check the resulting design - so far as resources permit - against the views of whichever interest-groups are relevant to the particular objective concerned.

The use of this process to meet the performance specification developed for the Newcastle example, in Design Sheet 5.1, is illustrated in the remainder of this Design Sheet.

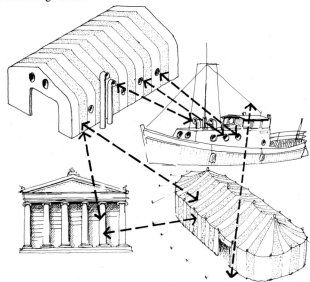

Regional Exhibition Centre, Newcastle, England

5.6: Contextual cues and use cues together

This Design sheet covers the complex situation which arises when we have to use contextual cues and use cues together, to satisfy the performance specification developed in Design Sheet 5.1.

In the process of choosing potential cues, as described in Design Sheets 5.3, 5.4 and 5.5, it will usually have become clear that some objectives can be met by a wide range of cues, whilst others can only be achieved with a very restricted range. To avoid abortive work when using the cues in design, it is important to begin by considering those objectives which can only be addressed with a *restricted* range of cues.

The first step, therefore, is to arrange the objectives in order, beginning with those which can only be addressed by a few cues, and ending with those which can be satisfied by the widest range. Once this has been done, start designing by selecting large-scale cues to satisfy the objective which has the most restricted range of appropriate cues. Then test whether the chosen cues are appropriate for the second objective, then the third, and so on; modifying the emerging design as necessary to meet as many of the objectives as possible. Then repeat this process with the small-scale cues, gradually building up the design in more and more detail, as illustrated in the example on this page and the next.

Cues \ Objectives	To seem an appropriate continuation of the quayside edge character formed by the exhibition centre, to as many groups as possible.	To seem appropriate as housing to the widest possible range of single people and young couples.	To seem appropriate as part of Newcastle city centre from across river, to as many groups as possible.	To seem appropriate as an outdoor leisure setting, to as many groups as possible.	Conclusion
Vertical rhythms	Exhibition centre verticals at 7m. centres (top). As yet, only verticals in flats are staircases, too widely spaced (bottom).	Make intermediate verticals: domestic bays, one per flat.	Procession of vertical bays could develop a classical character. No problem.	Bays and stairs imply places for lingering, not just a linear movement space. No problem.	**Project stairs, with vertical bays for each flat.**
Horizontal rhythms	Exhibition centre has each floor horizontally expressed. Use this as starting point.	Horizontal divisions will help express individual flats, reducing institutional image. No problem.	Bring out ground floor particularly strongly, as in local classical precedent above.	Domestic/classical cues quite appropriate as leisure setting: no problem.	**Express the different floors, bringing out the ground floor particularly strongly.**
Skylines	Skyline only seen across river at long range. Make mostly flat, as in exhibition centre.	Flat skyline potentially institutional. Encourage domestic associations close to, with gables over bays.	Bring out classical potential through grouping of bays and gables.	Quayside too close to see skyline: no problem.	**Make flat long-range skyline, with gables over pairs of bays.**

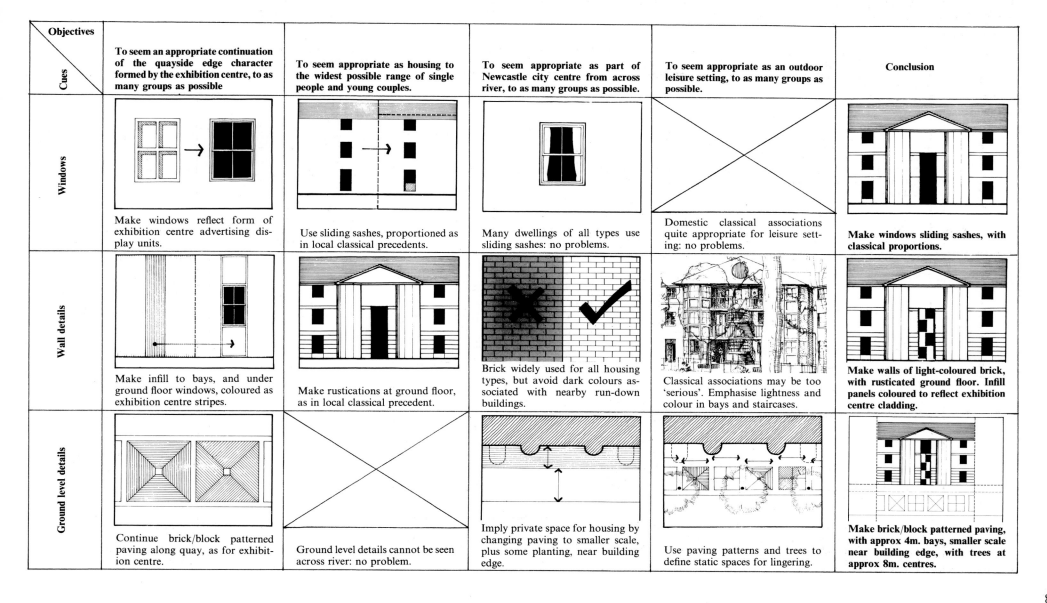

Objectives / Cues	To seem an appropriate continuation of the quayside edge character formed by the exhibition centre, to as many groups as possible	To seem appropriate as housing to the widest possible range of single people and young couples.	To seem appropriate as part of Newcastle city centre from across river, to as many groups as possible.	To seem appropriate as an outdoor leisure setting, to as many groups as possible.	Conclusion
Windows	Make windows reflect form of exhibition centre advertising display units.	Use sliding sashes, proportioned as in local classical precedents.	Many dwellings of all types use sliding sashes: no problems.	Domestic classical associations quite appropriate for leisure setting: no problems.	**Make windows sliding sashes, with classical proportions.**
Wall details	Make infill to bays, and under ground floor windows, coloured as exhibition centre stripes.	Make rustications at ground floor, as in local classical precedent.	Brick widely used for all housing types, but avoid dark colours associated with nearby run-down buildings.	Classical associations may be too 'serious'. Emphasise lightness and colour in bays and staircases.	**Make walls of light-coloured brick, with rusticated ground floor. Infill panels coloured to reflect exhibition centre cladding.**
Ground level details	Continue brick/block patterned paving along quay, as for exhibition centre.	Ground level details cannot be seen across river: no problem.	Imply private space for housing by changing paving to smaller scale, plus some planting, near building edge.	Use paving patterns and trees to define static spaces for lingering.	**Make brick/block patterned paving, with approx 4m. bays, smaller scale near building edge, with trees at approx 8m. centres.**

Quayside housing project, Newcastle, England

Chapter 6: Richness

Introduction
So far we have discussed ways of making major decisions about the layout and image of a building or outdoor place. But there is still room for manoeuvre in terms of detailed design. We must make the remaining decisions in ways which increase the variety of sense-experiences which users can enjoy. We call this quality *richness*.

Design for all senses
For most people, sight is the dominant sense. Most of the information we handle is channelled through our eyes, so a large part of this chapter is concerned with visual richness[1].

But richness is not a purely visual matter: other senses also have design implications:
- the sense of *motion*
- the sense of *smell*
- the sense of *hearing*
- the sense of *touch*

A starting point
Designers are mostly concerned with the fixed bits of places. For richness, we must design these to offer sensory choice. This implies designing so that people can choose different sense-experiences on different occasions. So we must begin by asking how users can choose different sense-experiences from a fixed environment.

How do users choose?
There are only two ways people can choose from different sense-experiences if the environment itself is fixed:
- by focussing their attention on *different sources* of sense-experience on different occasions.
- by *moving away* from one source towards another.

The effectiveness of each method depends on whether the sense concerned can be directed in a selective way, or whether it picks up information indiscriminately, from all sides at once. The senses vary from totally indiscriminate to highly selective, as shown in the diagram below:

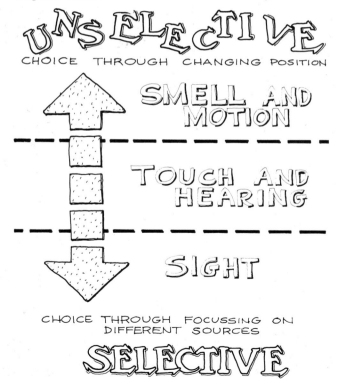

UNSELECTIVE

CHOICE THROUGH CHANGING POSITION

SMELL AND MOTION

TOUCH AND HEARING

SIGHT

CHOICE THROUGH FOCUSSING ON DIFFERENT SOURCES

SELECTIVE

The sense of motion
Choice of kinetic experience can only be gained through movement, so kinetic richness implies different possibilities for moving through a place. It is therefore mostly relevant to large spaces: outdoor places, and circulation routes within buildings.

The sense of smell
Because the sense of smell cannot be directed, choice of olfactory experience can only be achieved by moving away from one source towards another. So this is another potential for richness which is only possible in relatively large places.

The sense of hearing
We have only limited control over what we hear: the act of hearing itself is involuntary; though we can distinguish between sounds, concentrating on one rather than another. Aural richness *can* be achieved in small spaces, therefore, but only at the cost of imposing it on everyone there. This means it is best restricted to spaces large enough for people to escape altogether from the sound sources involved.

The sense of touch
Touch is both voluntary and involuntary in character: we can choose what we want to touch merely by moving a hand, but only by moving away can we avoid being touched by a breeze or a sunbeam. So richness of surface texture can be packed into the smallest of spaces, but variety of air movement and temperature should be reserved for large ones.

Designing for non-visual richness
Because current design thinking is almost entirely preoccupied with visual concerns, there is little useful theory about designing for non-visual richness. This is an area where research is urgently needed: in the meantime, all we can offer is a series of examples, as starting points for further exploration. These examples form the subject of Design Sheet 6.1.

The sense of sight
Vision is both the dominant sense in terms of information input, and the one most under our control. We have only to move our eyes to change what we look at.

This gives visual richness a double importance: it is what the rest of this chapter is about.

Why is visual richness a problem?

The visual monotony of many recent environments is now widely recognised, so designers' and patrons' attitudes are changing. But after fifty years of neglect, the principles of designing for visual richness have been forgotten. With no principles to go on, designers can only base their work on examples of richness from the past.

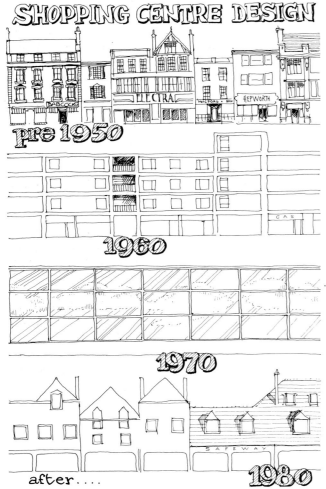

But we shall often need to design buildings whose images are *not* based on those of rich architecture from the past. The pastiche approach to richness is no help in that situation: we need a more solid basis from which to work.

The basis of visual richness

Visual richness depends on the presence of visual contrasts in the surfaces concerned. The most effective means of achieving such contrasts depends on two main factors:

- the orientation of the surface concerned
- the likely positions from which it will be viewed

The design implications of these factors are discussed in Design Sheet 6.2.

Using contrasts to achieve richness

By this stage in design, the surfaces of the scheme already contain visual contrasts, formed by the cues used for achieving visual appropriateness in Chapter 5. It is these cues which must be developed to gain further richness if necessary.

The richness already achieved in Chapter 5 depends on the *number* of visual elements present in each surface, and on the *relationships* between them.

For example, if a particular surface consists of only one element, as illustrated below, it contains no choice of things to be looked at, and therefore no visual richness.

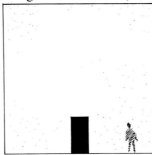 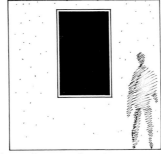

As the number of elements in a given surface increases, so does richness. By the time the surface contains about five elements[2], there is plenty of choice of things to look at, so the surface seems rich, as sketched below.

But when the number of elements exceeds a certain level, the various elements begin to be read together, as a single pattern or super-element. When this happens, richness of experience is reduced. As a rough guide, this is likely to happen when the number of elements concerned exceeds about nine, as sketched below.

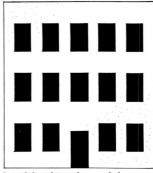 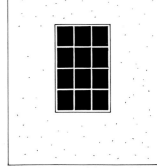

In this situation, richness can be increased by making larger-scale subdivisions of the surface concerned, so that the elements will no longer all be read together, but are separated into groups of between five and nine: below five, richness is low because there is insufficient choice of things to look at; while above nine, the whole ensemble will still be read as a single super-element, with no visual choice.

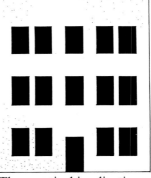 

The practical implications of this rough-and-ready rule vary according to two key factors:

- the likely range of distances from which the surface concerned will be viewed.
- the length of time during which each view will be experienced.

The next step in designing for richness is therefore to assess these two factors for each of the surfaces concerned. This is covered in Design Sheet 6.3.

Viewing distance

The range of likely viewing distances affects the range of scales at which richness must be considered. Where the surface will be seen at long range, large-scale richness is necessary; whilst at close range, richness must be achieved by small-scale elements and subdivisions. So to maintain richness from long-range to close-range we need a *hierarchy* of elements from large-scale to small-scale. This topic is explored in Design Sheet 6.4.

Richness, at a variety of scales, is maintained from long-range to close-range: Old Indian Institute building, Oxford, England.

Viewing time

Where people are likely to view a particular surface from a given position for a long time, it is important that the surface concerned should continue to seem rich for as long a period as possible. Ways of achieving this are discussed in Design Sheet 6.5.

The cost of richness

Richness does not always cost more than plainness. Surprisingly often, plainness requires an additional skin to be added; or visual simplicity is achieved by complex and expensive secret fixings.

In practice, though, close-range richness often does cost more. This is not surprising: neither designers nor builders are used to it nowadays, and it takes longer to design. Extra construction costs will be minimised by our policy of designing richness only for relevant viewing positions. In this way, we shall ensure that any extra money is spent in a cost-effective way. But it is important that the techniques and materials we use are also as cost-effective as possible.

Techniques and materials

In the past, close-range richness was feasible only because craftsmen were paid very little. Happily, this is no longer the case: for most buildings, we must find ways of enriching places which take advantage of modern production techniques, and accept modern labour costs. Some feasible approaches are listed below:

- when using mass-produced components, consider the *range* available, rather than arbitrarily repeating a single element.
- consider *revealing* construction and fixings, rather than hiding them.
- for close-range richness, use materials with *inherent surface variety.*
- consider how richness might be increased by harnessing the *builder's* initiative.
- recycle craftsmanship: *re-use* the richness from the past, which we can no longer afford.

The Design Sheets which follow show examples of these approaches.

Design implications

How to achieve richness

1. **Take the detailed design from Chapter 5 as the basis for developing further richness.**

2. **Decide which locations have potential for non-visual richness, and design for kinetic experience, smell, hearing and touch (Design Sheet 6.1)**

3. **Analyse the various surfaces of the scheme, to assess the most appropriate strategies for achieving visual contrasts (Design Sheet 6.2)**

4. **Analyse likely viewing distances and times for each surface, and the relative numbers of people concerned (Design Sheet 6.3)**

5. **Develop each of the surfaces designed in Chapter 5; increasing its richness, if necessary, over its full range of viewing distances (Design Sheet 6.4)**

6. **Develop extra richness for those surfaces which are likely to be viewed for long periods (Design Sheet 6.5)**

7. **Check feasibility in terms of materials and techniques, and amend design if necessary.**

6.1: Non-visual richness

Current design thinking is almost entirely preoccupied with visual concerns: there are very few places designed specifically for the non-visual senses, and still less theory about how such places *could* be designed. This is a topic which urgently needs investigation: in the meantime, all we can do is offer some examples, as a starting point for further exploration.

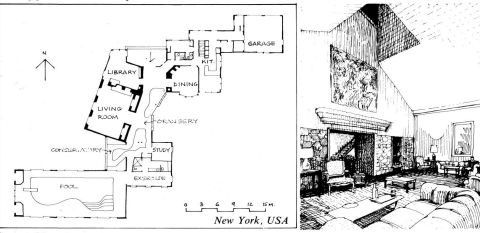

New York, USA

Sense of hearing

In this house by Charles Moore, the floor finishes are designed to make different sounds underfoot, whilst the volumes of the various internal spaces are designed to provide a variety of reverberation times. The result is a rich acoustic environment throughout the house as a whole.[34]

Oxford, England

Sense of touch

Portland Square, by Lawrence Halprin, uses water to provide a rich range of tactile experiences. This is also achieved in Helen Teague's project for an infants' school, with its variety of different floor textures and door handles[5].

Sense of smell

A large formal herb garden seen through a narrow doorway cut into a high yew hedge. The beds are filled with highly scented plants, whose aroma is concentrated within the wind-free hedged enclosure[6]. In urban places, remember the potential of cafés, bakeries and the like: give them the opportunity of opening up to the space outside.

Paris, France

Sense of motion

The Centre Pompidou, by Piano and Rogers, uses escalators to provide a variety of movement sensations; experienced in relationship both to close-up parts of the building itself, and to distant city views[7].

6.2: Visual contrasts

Visual events depend on visual contrasts, which can be created by differences of colour or tone on a two-dimensional surface, or by three-dimensional variations of the surface itself. The relative effectiveness of these approaches depends on two main factors:
- the orientation of the surface concerned
- the likely positions from which it will be seen.

Rome, Italy **1**

Vienna, Austria **2**

Keble College, Oxford, England **3**

Use contrasts of colour or tone on floor or ground surfaces which have to be flat (1); or on surfaces or materials which are unsuitable for three-dimensional modelling (2); or where a flat surface is all that can be afforded (3).

Khajuraho, India **4**

Hollyhock House, Los Angeles, Ca., USA: Frank Lloyd Wright **5**

Hoover building, London, England **6**

Use three-dimensional variations where strong light will sharpen contrasts (4); or where a self-coloured material lacks strong colour contrasts, or where a material is unsuitable for colouring (5); or in addition to colour in especially significant areas (6).

7

8

9

Horizontal projections are important to richness when people are likely to be walking parallel to the building, and near to its surface, as in street architecture (7). Richness at high level, from close to, also depends on such projections (8). From further away, high level richness depends most on vertical projections (9), usually visible only in outline or silhouette.

6.3: Viewing distances, numbers and times

Appropriate decisions about visual richness must take three main factors into account:
- the range of distances from which the various parts of the scheme can be seen.
- the relative numbers of people likely to see the building from each different viewing position.
- the length of time during which each view will be experienced.

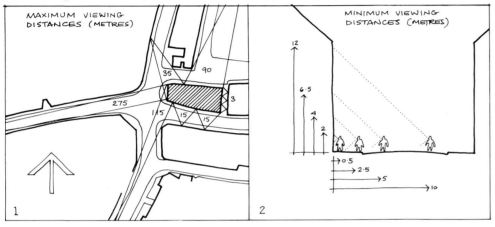

To contribute to visual richness, elements have to be visible. What users can see depends partly on how far away from the building they are. So begin by analysing the range of distances from which each part of the building can be seen. Remember the viewpoints of the occupants of adjoining buildings (1,2).

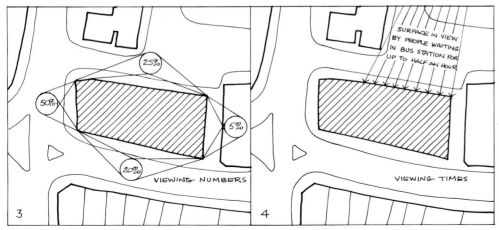

To make sure that investment in richness is used to the best effect, it is important to know the relative numbers of people likely to experience each of the views noted so far (3). Note also any surfaces which may be viewed for long periods, for example by people waiting for buses; or, at closer range, by people waiting to be admitted at entrances (4).

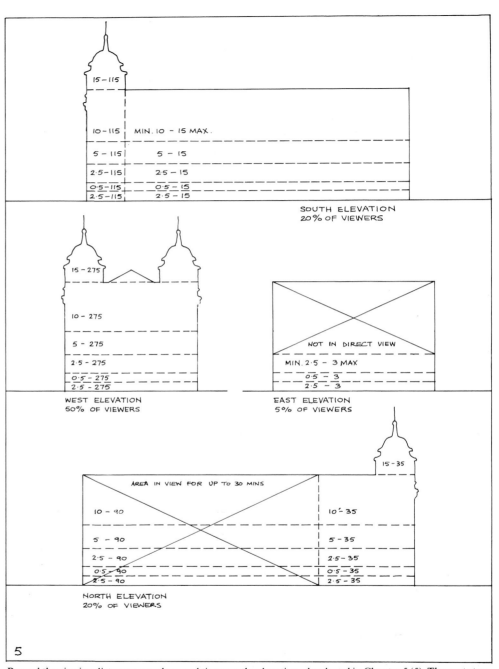

Record the viewing distances, numbers and times on the elevations developed in Chapter 5 (5). The next step is to design appropriate levels of richness for the various situations thus defined. This is covered in Design Sheets 6.4 and 6.5.

6.4: Implications of viewing distance

In Design Sheet 6.3, maximum and minimum viewing distances were noted for each part of the building. This information must now be used to design appropriate levels of richness for each of the areas concerned.

Begin by considering the maximum viewing distance for each surface concerned. Draw the surface to the size it will appear when viewed from that distance (1).

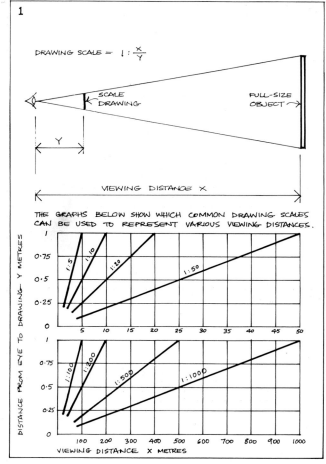

DRAWING SCALE = $1 : \frac{X}{Y}$

SCALE DRAWING

FULL-SIZE OBJECT

VIEWING DISTANCE X

THE GRAPHS BELOW SHOW WHICH COMMON DRAWING SCALES CAN BE USED TO REPRESENT VARIOUS VIEWING DISTANCES.

DISTANCE FROM EYE TO DRAWING Y METRES

VIEWING DISTANCE X METRES

If, when drawn to this size, the surface shows less than five distinct visual elements, then redesign it to include more, up to a maximum of nine (2). If it has more than nine, then redesign it to group some of them together, to read as having between five and nine elements (3).

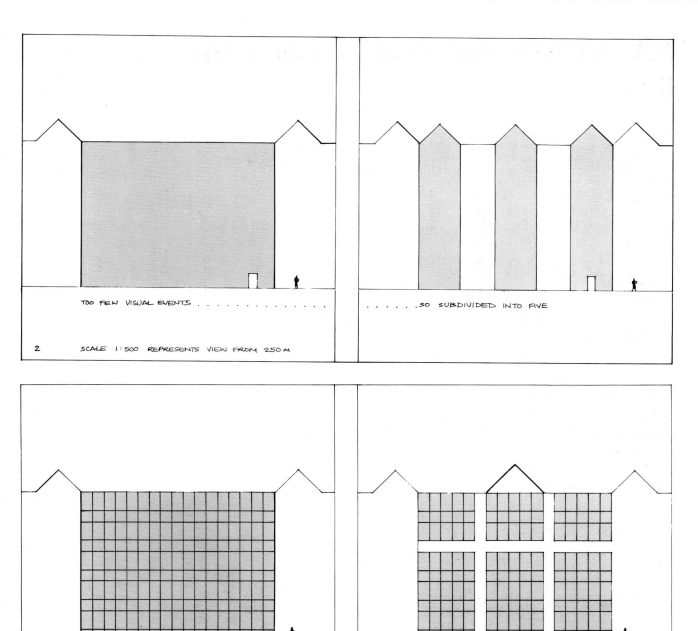

TOO FEW VISUAL EVENTS . . . .

. . . . . . .SO SUBDIVIDED INTO FIVE

2   SCALE 1:500   REPRESENTS VIEW FROM 250 M

TOO MANY SIMILAR VISUAL EVENTS. . . . .

. . . . SO GROUPED TOGETHER

3   SCALE 1:500   REPRESENTS VIEW FROM 250 M.

Next draw the elevation to about three times the previous scale, showing all the elements which are visible at this scale. Check how many elements are revealed.

If less than five, subdivide into between five and nine sub-events. If more than nine, then group some of them together (4).

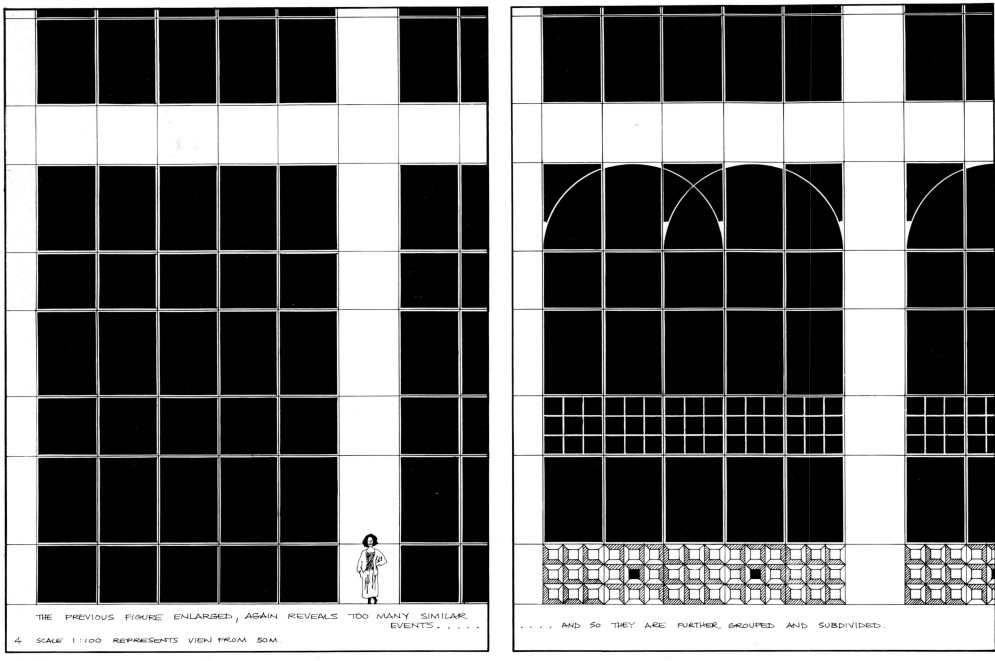

THE PREVIOUS FIGURE ENLARGED, AGAIN REVEALS TOO MANY SIMILAR EVENTS.....

4 SCALE 1:100 REPRESENTS VIEW FROM 50M.

.... AND SO THEY ARE FURTHER GROUPED AND SUBDIVIDED.

Continue to redraw the surface concerned, to approximately three times the previous scale (5). At each stage, check the subdivision of elements as previously described. Continue this process until you reach a scale appropriate to the shortest viewing distance.

Finally, check the effects of viewing angle on richness: if many visual events are masked, add more in the form of projections, as described in Design Sheet 6.2, but again working within the vocabulary of cues developed in Chapter 5 where possible.

PART OF THE PREVIOUS FIGURE ENLARGED WITH VARIOUS SMALLER VISUAL EVENTS ADDED.

5 SCALE 1 : 20 REPRESENTS VIEW FROM 10M.

6.5: Implications of viewing time

Where people are likely to view a particular surface from a given position for a long time, as pin-pointed in Design Sheet 6.3, then the design from Sheet 6.4 should be developed still further. It is important that the surface concerned should continue to seem rich over a long period. This can be achieved in three main ways:
- through greater visual complexity.
- through visual riddles.
- through interpretation.

Natural History Museum, London, England.   *NUK Library, Ljubljana, Jugoslavia.*

These examples (1) show surfaces designed with a high level of visual complexity, within which many alternative patterns can be discovered over time.

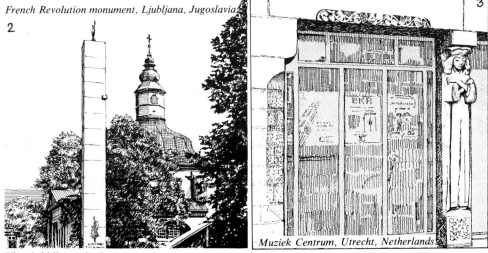

French Revolution monument, Ljubljana, Jugoslavia. 2   3   *Muziek Centrum, Utrecht, Netherlands.*

Visual riddles, as in these examples, engage people's creative imagination in making sense of them. Why is that one stone out of place? (2). And why does that old sculpture appear in the middle of this window? (3).

Michelin building, London, England. 4   *Brussels, Belgium.*

In these examples (4) the surfaces are made to yield more information by the addition of extra interpretative material.

5

This example (5) shows the elevation designed in Sheet 6.4, developed by using all three of the approaches outlined above.

Chapter 7: Personalisation

Introduction

In the previous chapters, we have covered ways of achieving the qualities which support the responsiveness of the environment itself, as distinct from the political and economic processes by which it is produced. This is not because we do not value the 'public participation' approach: from our point of view, it is highly desirable. But we have already explained that even with the highest level of public participation, most people will still have to live and work in places designed by others.[1]

It is therefore especially important to make it possible for users to *personalise* these existing environments: this is the only way most people can achieve an environment which bears the stamp of their own tastes and values. Paradoxically, this calls for considerable effort from the place's original designer. This chapter is about how to apply that effort in the most effective way.

Personalisation and legibility

There is a secondary reason for supporting personalisation: it makes clearer a place's pattern of activities. This is particularly valuable in robust environments, accommodating a wide variety of uses, changing over time. By encouraging each user to dress the building differently, personalisation can make each use explicit.

Current trends

Personalisation seems to be increasing nowadays, partly because of an ever-increasing range of cheap ways of changing buildings' external appearance.

Under certain circumstances, the combined effects of these changes become a political issue: the kind of transformation shown below generates heated planning debates about control versus individual choice.[2]

Grimsby Borough Council (1976)

Grimsby Borough Council (1976)

This is only a problem where personalisation is not thought through as an integral part of the original scheme: we shall return to this later in the chapter. First we must explore the process of personalisation itself, and how it can be encouraged.

Types of personalisation

Users personalise in two ways:
- to improve *practical facilities*.
- to change the *image* of a place.

Chapter 4 has already covered ways of enabling users to adapt buildings' practical facilities, so in this chapter we shall concentrate on personalising the *image* of a place.

Why personalise images?

People personalise a building's image for two main reasons:
- as an affirmation of their own tastes and values: *affirmative* personalisation
- because they perceive its existing image as inappropriate: *remedial* personalisation

From our standpoint, affirmative personalisation must clearly be supported. Sometimes, designers deliberately encourage remedial personalisation, designing inappropriate images to *incite* personalisation.[3]

This is a form of architectural coercion: as we have explained in Chapter 5, people only develop truly participatory relationships with places they like. In this chapter, therefore, we shall concentrate on affirmative personalisation.

Constraints on personalisation

Personalisation is affected by three main factors:
- tenure
- building type
- technology

Tenure

Personalisation is unlikely to happen unless the user of a place has a claim to its occupation, whether by custom or legal fiat. The way this claim is controlled - particularly by the building's *owner* - has radical effects on whether and how personalisation takes place. The balance of power between user and owner is set by the tenure system:

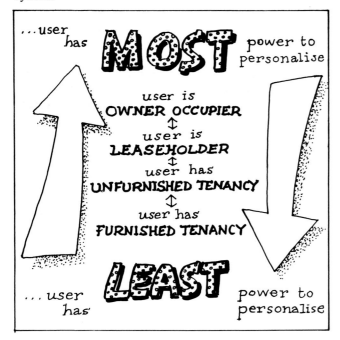

...user has **MOST** power to personalise

user is
OWNER OCCUPIER
⇕
user is
LEASEHOLDER
⇕
user has
UNFURNISHED TENANCY
⇕
user has
FURNISHED TENANCY

...user has **LEAST** power to personalise

Tenure affects two key aspects of personalisation:
- the money spent on it
- its permanence

Though designers will know the initial tenure of their projects, remember that the robust buildings we advocate will probably *change* their tenure system over time. Since it costs little more to encourage personalisation, architects should allow for the full *range* of tenure, except for specialised building types where tenure is unlikely to change.

Building type

People mainly personalise places they regularly use for long periods: in practice, homes and workplaces. Nearly all buildings, at least in part, contain either homes or workplaces, or may do so in the future. Most buildings, therefore, should be designed to encourage personalisation.

Though most buildings should encourage personalisation, all but the smallest have public areas which will probably not be personalised because nobody stays there long enough. These places are often the areas of most public significance, and their lack of personalisation will call for extra *richness*, as covered in Chapter 6.

Technology

Supporting personalisation includes making it physically easy. This means that the technology of the design should be well-matched to the expertise of the likely users. Since expertise is hard to predict, it is best to use materials and techniques which unskilled people can easily master, at least where personalisation is most likely.

Where does personalisation happen?

In personalising a place, users are both confirming their tastes and values to themselves, and communicating them to others. The former occurs mostly *inside* a user's space, and the latter *across its boundary*, real or implied. This boundary separates the user's private domain from the public realm: it enables us to make the important distinction between *private* and *public* personalisation.

Private personalisation

The physical elements supporting personalisation *within* a space consist of internal surfaces and focal elements. These are covered in Design Sheet 7.1.

Public personalisation

Some personalisation communicates *across* the private/public boundary, affecting the public realm. This mostly happens at physical gaps in the boundary:
- thresholds (covered in Design Sheet 7.2)
- windows (covered in Design Sheet 7.3)

More overtly public in intention is personalisation of the outer surface of the boundary itself. This is covered in Design Sheet 7.4.

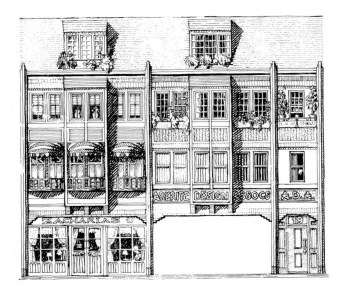

Public impact

If a building is not designed specifically to accept it, a high level of personalisation may be detrimental to its public role; eroding the balance between pattern and variety sought in Chapter 6. Personalisation may overwhelm too fragile a pattern altogether.

If this happens it becomes a political issue, because *private* actions erode the quality of the *public* realm. It does *not* mean we should repress personalisation. Rather it calls for buildings which can accept it without degenerating into chaos.

Patterns of personalisation

Personalisation is not random. People personalise only the space *they* control, so - as the above picture shows - patterns of personalisation reflect patterns of tenure. These are predictable: even with highly robust buildings it is not difficult to establish the most likely possibilities. Once this is done, the probable effects of personalisation can roughly be estimated, to see whether they are likely to disrupt the qualities already designed into the scheme. Design Sheet 7.5 explores how this can be done, and how the design might be modified to encourage personalisation without destroying either visual appropriateness or richness.

Design implications

How to encourage personalisation

1. **Take the design from Chapter 6 as the starting point for developing the project to support personalisation.**

2. **Develop the detailed design of internal surfaces (Design Sheet 7.1)**

3. **Develop the detailed design of thresholds, both internal and external (Design Sheet 7.2)**

4. **Develop the detailed design of windows (Design Sheet 7.3)**

5. **Develop the detailed design of external surfaces. Assess likely effects of publicly-visible personalisation, and amend the design if necessary (Design Sheet 7.4)**

7.1: Internal walls

Internal walls can be personalised in two main ways:
- by using them as display settings
- by decorating their surfaces

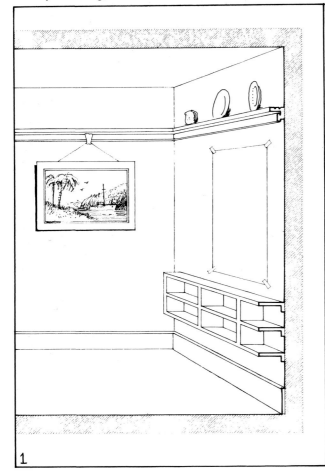

For convenient display, the wall should be easy to fix things to. The core material should be soft enough to take 'hammer and screwdriver' fixings, but hard enough to carry extensive shelf units. Smooth flat surfaces are best for wallpapering. To make this easier, opportunities for picking out details in different colours - which require articulated surfaces - are best at tops and bottoms of walls. Picture rails make it easy to hang things and - if wide enough - can be used as display shelves.

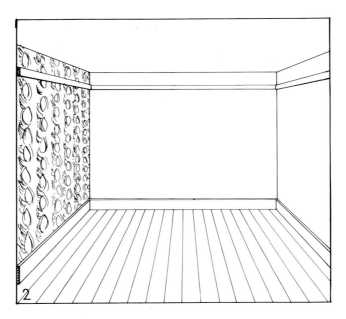

Picture rails can also define an ambiguous zone between ceiling and wall, which can be painted in with either; thus providing easy opportunities to change the apparent proportions of the room (2,3).

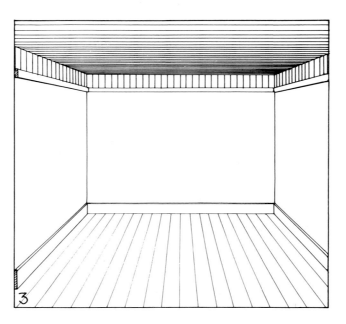

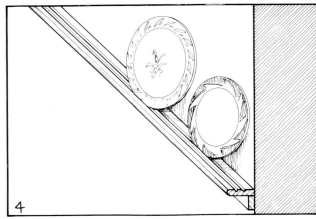

Some low-level projections, such as radiators, may be unavoidable for practical reasons. They should be flat-topped so that things can be stood on them. Grooved tops allow plates and photos to lean without special supports (4) but are still OK for books. Where fireplaces occur, they have special significance for display; lending the symbolic importance of the hearth to objects placed on them (5).

7.2: Thresholds

A threshold is a physical link between different people's domains. It is therefore a key area for the display of a person's or group's own values.

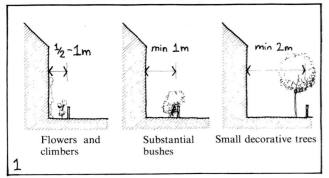

In housing, front gardens offer the largest scale of threshold space, so include them where possible (1). Where this cannot be achieved, small unit paving at the building edge can easily be adapted for planting (2).

Porches offer scope for 'picking out' (3) and make potential spaces to shelter objects on display (4): if you omit them, leave clear wall spaces around the front door to allow later additions (5). Walls adjacent to front doors are also frequently personalised, even in non-residential buildings, so fixings should be made easy here.

6 *Old people's home, Amsterdam, Netherlands Herman Hertzberger*

Threshold spaces to front doors inside buildings can also support personalisation (6). The more private thresholds to individuals' rooms will be more squeezed for space, but display over the door may be possible (7). The door construction itself is also important: exposed construction lends itself to picking out, and makes it obvious where to fix embellishments.

7.3: Windows

Like thresholds, windows are important for personalisation because they form physical links between private and public worlds. They offer three main kinds of potential:
- for display through them
- for external display associated with them
- for alterations to the windows themselves.

 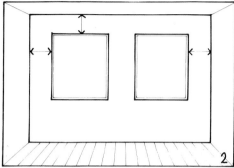

There are three main kinds of display from inside: objects stood on the cill, things suspended from the head and curtains. Cills should be of hard rot-proof material, wide enough to take plant-pots etc. (1). Heads and jambs should take fixings easily. Internally, leave space over the windows for pelmets, and at the sides so that curtains can be drawn right back (2).

 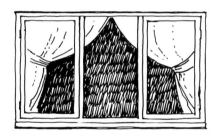

Small objects can look lost in a big pane of glass. Glazing bars can be used to frame them appropriately, but remember that people often display a single main object centrally placed: avoid obscuring it with a central glazing bar (3). Curtain displays are almost invariably symmetrical, and look best against a symmetrical arrangement of glazing bars (4).

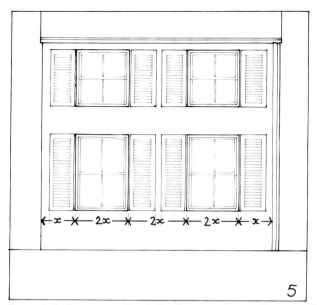 

External displays are usually window boxes or shutters. Shutters affect window spacing (5). Window boxes affect how the windows will open. They must open for cleaning and plant maintenance, but should not open outwards to avoid damaging plants - if these are to be visible from inside - or inwards because of internal cill displays. Sliding windows - vertical or horizontal - are therefore best. Building in fixings for window boxes can remind users of the possibility, and add richness even when not used (6).

Despite their other advantages, glazing bars limit painted displays on the windows themselves (7). Window frames which need painting give opportunities for personalisation through colour selection: as with doors, the more complex they are, the greater the opportunities for 'picking out'(8). But do not take this to extremes: limit the areas requiring maintenance and make it easy to carry out, on upper floors, from inside.

7.4: External surfaces

External surfaces should be designed to encourage personalisation. But when the surfaces are publicly visible, they should also be designed so that personalisation will not destroy the visual appropriateness and richness developed in Chapters 5 and 6.

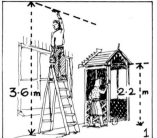  

Some parts of a building's external surfaces are more easily accessible than others (1,2,3): it is here that opportunities for personalisation are most available to the ordinary user.

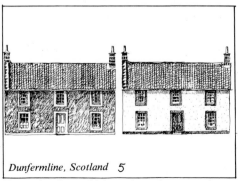

Dunfermline, Scotland 5

YES DARLING.... I KNOW ITS YOUR FAVOURITE....

BUT IT WILL HAVE TO COME OUT TO PAINT THE BOARDING.

External surfaces which require maintenance give automatic opportunities for personalisation (4). These are increased if the areas concerned are articulated to support picking-out (5). But balance the proportion of such surfaces against the costs of maintenance. And remember that large wall areas which need maintenance will interfere with opportunities for personalising the surface with climbing plants (6).

On a drawing, simulate the types and degrees of personalisation which you think people are likely to carry out. Check whether any of the design characteristics which are important for visual appropriateness might be obliterated (7). If so, reinforce them with as many design features as possible. Even if some features are changed, this reinforcement will increase the likelihood that enough will remain to make the visual appropriateness read through.

Where key features are positioned *between* different users, even the fiercest personalisation is unlikely to obliterate them (8).

Chapter 8:
Putting it all together

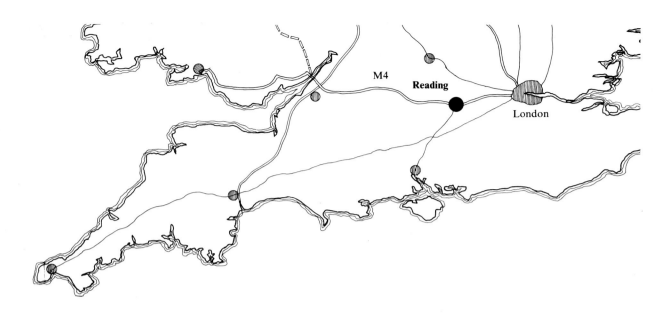

Purpose

This chapter works through the Design Sheets, in the order followed in the previous sections of the book; showing how our approach can be used to design a complex project on a large city-centre site.

The site

Covering an area of 5.6 hectares, the site lies near the centre of Reading: a thriving town with a population of about 150,000, some 30 miles west of London, along the M4 motorway to Bristol.

The site adjoins the southern edge of the town's commercial centre. Its southern boundary is formed by a major inner distribution road, giving good road linkage to the rest of Reading, and to other towns nearby. Bridge Street - which links this distribution road to the heart of the town centre - bisects the site itself.

Historical background

For the last two centuries the site - once an area of marshland adjoining the River Kennet - has been used by a brewery. Originally, the brewers were attracted to the area by the ready availability of water; necessary both for the production process and for transport. Neither of these factors is important nowadays, and the brewery has now moved its production areas to another site on the outskirts of the town; thus releasing its city-centre site for redevelopment.

As yet, however, the site has not been cleared: it still contains a variety of buildings, mostly from the nineteenth century. Three of them - a house attributed to Sir John Soane, a fine maltings and a former stable block - are listed by the Department of the Environment as having special architectural and historical importance.

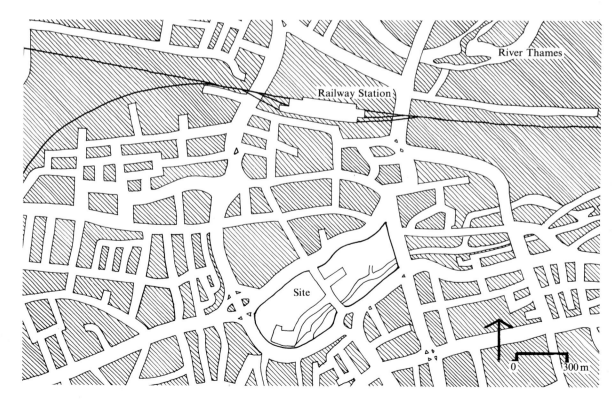

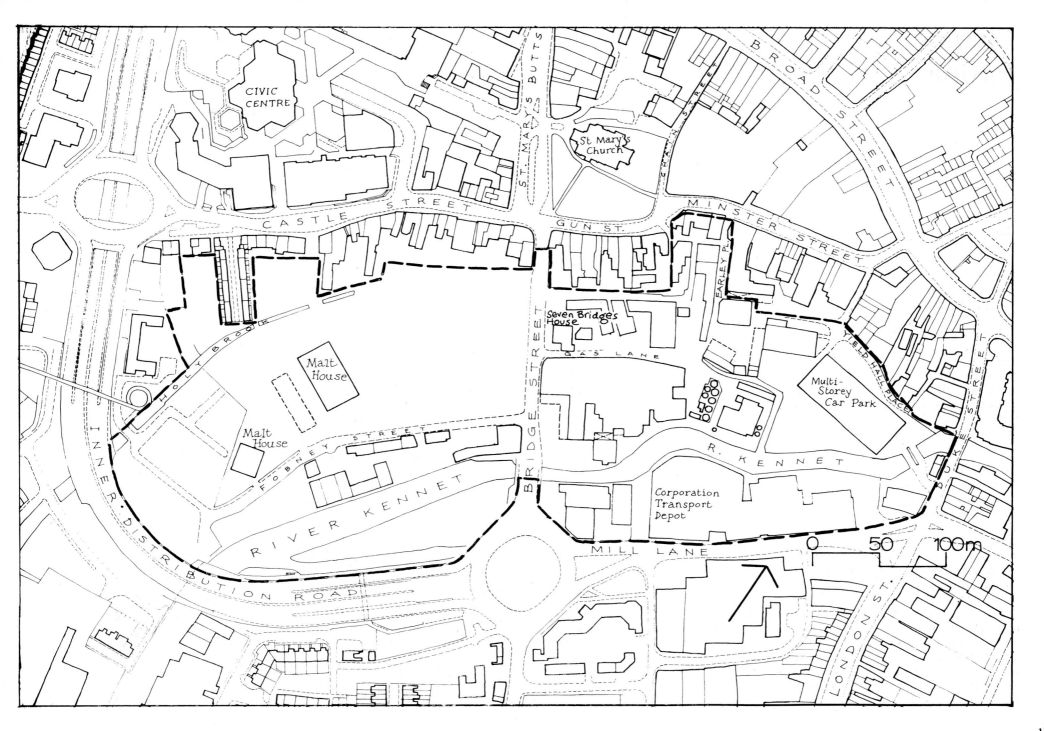

CIVIC
CENTRE

ST MARY'S BUTTS

St Mary's
Church

CHAIN STREET

BROAD STREET

CASTLE STREET

GUN ST.

MINSTER STREET

EARLEY PL.

YIELD HALL PLACE

Seven Bridges
House

HOLY BROOK

Malt
House

GAS LANE

Multi-
Storey
Car Park

Malt
House

FOBNEY STREET

BRIDGE STREET

R. KENNET

Corporation
Transport
Depot

DUKE STREET

RIVER KENNET

INNER DISTRIBUTION ROAD

MILL LANE

0   50   100m

LONDON ST.

107

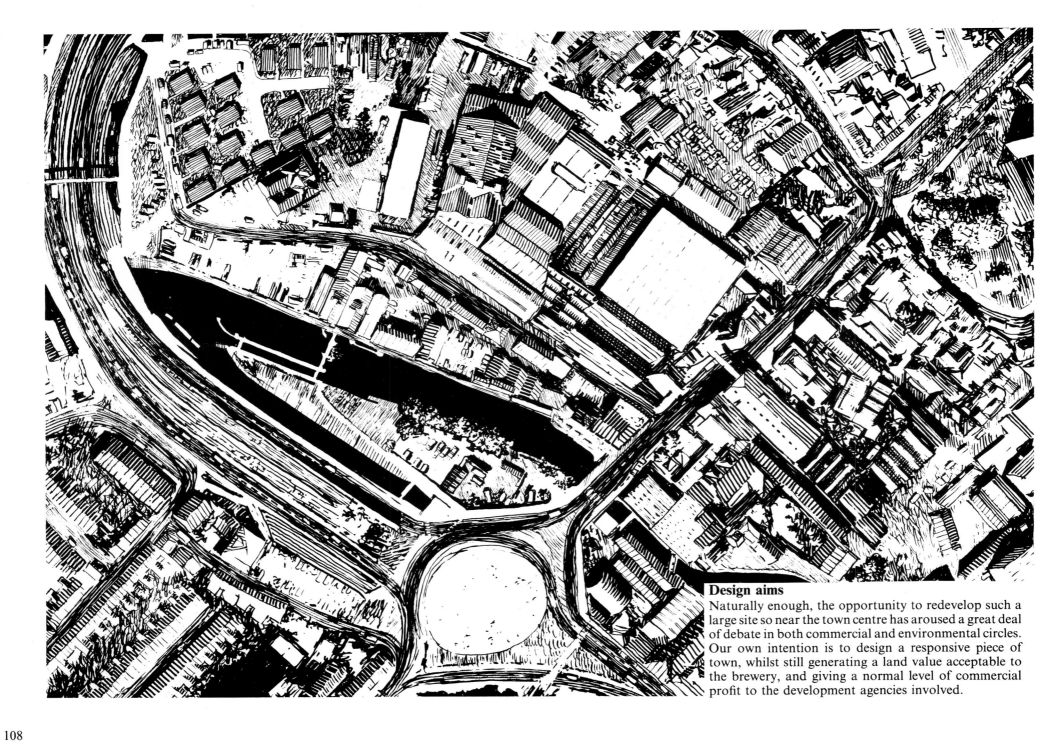

Design aims
Naturally enough, the opportunity to redevelop such a large site so near the town centre has aroused a great deal of debate in both commercial and environmental circles. Our own intention is to design a responsive piece of town, whilst still generating a land value acceptable to the brewery, and giving a normal level of commercial profit to the development agencies involved.

The starting point for a permeable scheme is the existing system of links into and through the site from the surrounding areas. This sheet analyses links between the site and the city as a whole, and links with the immediate local surroundings.

Figure 1 explores the relative importance, in both city-wide and local terms, of all the site's potential links. Clearly it is crucial to connect any new scheme into those links which achieve a high score in this analysis; though in the case of this particular site, there is at this stage no reason not to connect to *all* the existing links. It is particularly important to maintain and develop the pedestrian connections through the south and west edges of the site, to prevent the Inner Distribution Road forming an impermeable barrier.

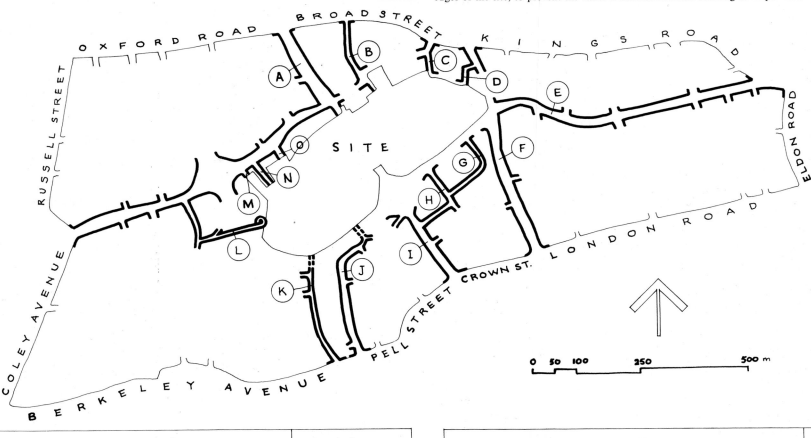

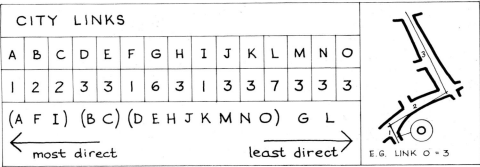

CITY LINKS

A	B	C	D	E	F	G	H	I	J	K	L	M	N	O
1	2	2	3	3	1	6	3	1	3	3	7	3	3	3

(A F I) (B C) (D E H J K M N O) G L

← most direct   least direct →

E.G. LINK O = 3

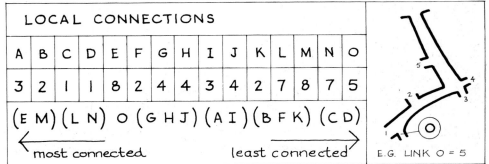

LOCAL CONNECTIONS

A	B	C	D	E	F	G	H	I	J	K	L	M	N	O
3	2	1	1	8	2	4	4	3	4	2	7	8	7	5

(E M) (L N) O (G H J) (A I) (B F K) (C D)

← most connected   least connected →

E.G. LINK O = 5

Links to the site: current situation

This sheet and the next illustrate the point at which each link joins the site itself, noting any significant implications from the previous analysis. The letters which denote each link are the same as those used in the analysis.

A. St Mary's Butts
The major vehicular link to north Reading and to the northern section of the Inner Distribution Road.

B. Chain Street
Typical of the narrow streets running north-south through the central shopping area. The most direct pedestrian connection from the site to the shops in Broad Street.

C. Yield Hall Lane
A narrow street joining Minster Street close to its connection with shops in Broad Street and King Street. Currently used as access to a multi-storey car park on our site.

D. Thorn Lane
A narrow link to the existing shopping area. Less direct than B or C.

E. Queens Road
The major vehicular link to east Reading, continuing the southern section of the Inner Distribution Road. Well connected to the local area to the east of the site.

F. London Street
A major vehicle link to south Reading.

G. Crossland Road
An alley which bends to join Letcombe Street, for service access only. The least important link to the site.

H. Letcombe Street
Joins Southampton Street, but also useful for local connections.

I. Southampton Street
As the name implies, a major vehicular link to the south.

J. Subway
Runs under the Inner Distribution Road to Katesgrove Lane. An important pedestrian link to housing south of the site. At present, does not extend across the river into the site.

K. Towpath
Runs under the Inner Distribution Road. Eventually links to housing areas to south and west, so should remain as a useful pedestrian link: it is particularly important to reinforce the limited connections across the Inner Distribution Road.

L. Footbridge
Runs over the Inner Distribution Road, to Coley Place. Again, this link is important since it is now the only pedestrian connection to the large housing areas to the west of the site.

M. Towrite's Yard
Vehicle access to Castle Street, and thence to the Inner Distribution Road. The only potential vehicular access to west Reading. Well connected to the local area west of the site.

N. Vachel's Almshouses
Pedestrian access to Castle Street.

O. Gap through Castle Street
A minor link for vehicles, but useful as a pedestrian route to and from the nearby Civic Centre.

The street/block system

Having analysed the system of links, the next step is to connect them to a preliminary street/block structure within the site. Since the uses to be included in the scheme have not yet been established, it is not possible to decide street widths and junction designs, nor to check block sizes.

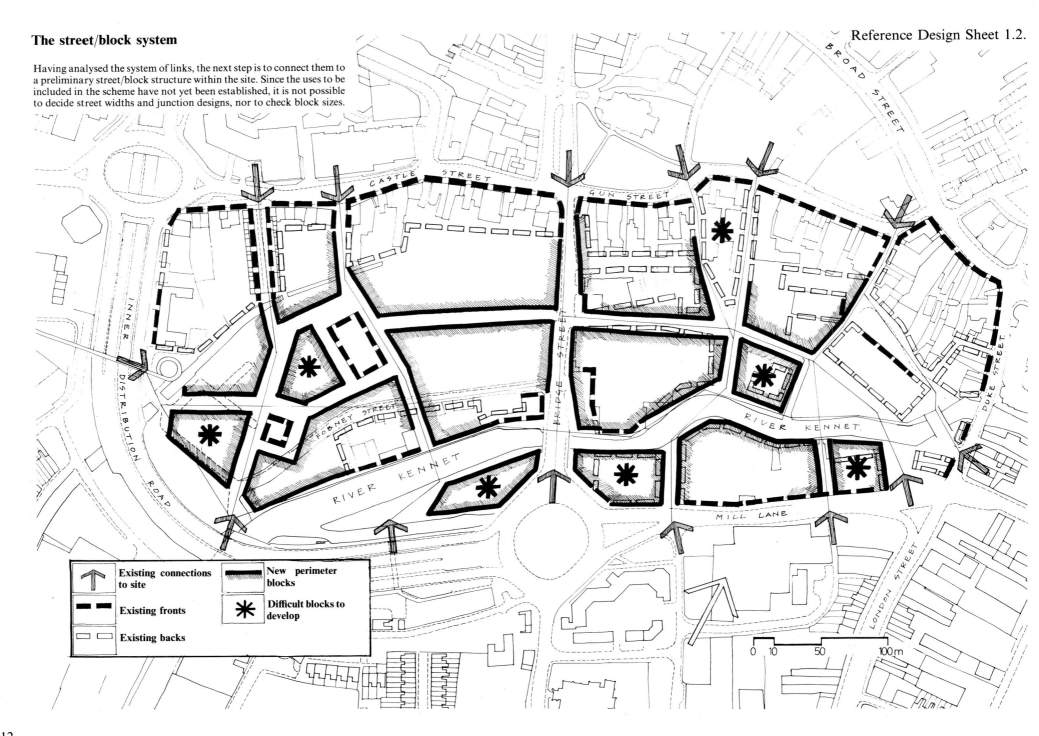

Uses and their compatibility

Having established a preliminary street/block system, the next step is to consider the variety of uses to be included in the scheme. To investigate demand, we consulted estate agents, and representatives of the local authority and various local organisations. The results of these enquiries suggested demand for the following uses:

- shopping.
- offices.
- residential (from small flats to 3-bedroom houses).
- indoor leisure facilities.
- TV studio/community theatre.
- canal basin.
- pubs.
- car parking.

Data from the various interviews were recorded on the pro-forma illustrated on the right. The likely interactions between the various uses were explored in the matrix below, which suggested the strategic location of uses sketched out on the next page.

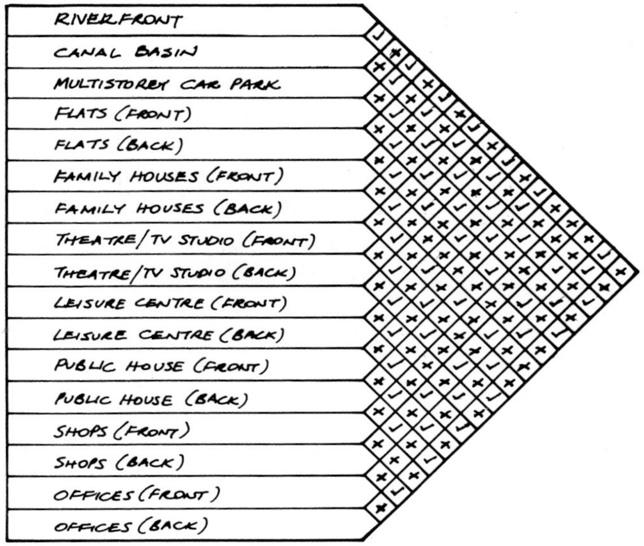

Site	COURAGE'S BREWERY SITE, READING
Interview with:	Peter Massif, Gibson Eley and Co.
Use	Offices
Possible developers	Major national developers, selling on to major financial institutions.
Use existing buildings?	No
Ancillary supports	See notes below
Demand: max/min areas	Min. individual building size 1000 m.s. Max. total area 35,000 m.s. (to include approx 10,000 m.s. for Courage's own use).
Rents	£12.50 per square foot
Yields	See notes below
Negative interactions	Do not combine with residential on top or underneath. If part has to be over shops, then yield will increase to 6½%. But avoid this as far as possible: such units may be very hard to let.
Planning controls	Local Authority expects high level of planning gain from the scheme. Situation is complex: contact officers directly as soon as possible.
Subsidies	None available
Notes	Ancillary supports Car parking under direct control of office tenants: not shared space in car park intended for shopping. Also important to achieve 'high quality of internal setting', to compete effectively with out-of-town campus offices. Riverside location therefore advantageous. Must be air-conditioned. Yield Yields vary between 4½% for offices prelet to a tenant with a really good covenant, up to about 7% if sold on unlet. Use 6% for prelim. calculations.

113

Strategic location of uses

Having analysed the likely interactions between the various uses proposed for the scheme, the next step is the strategic allocation of uses to approximate locations in the street/block system.

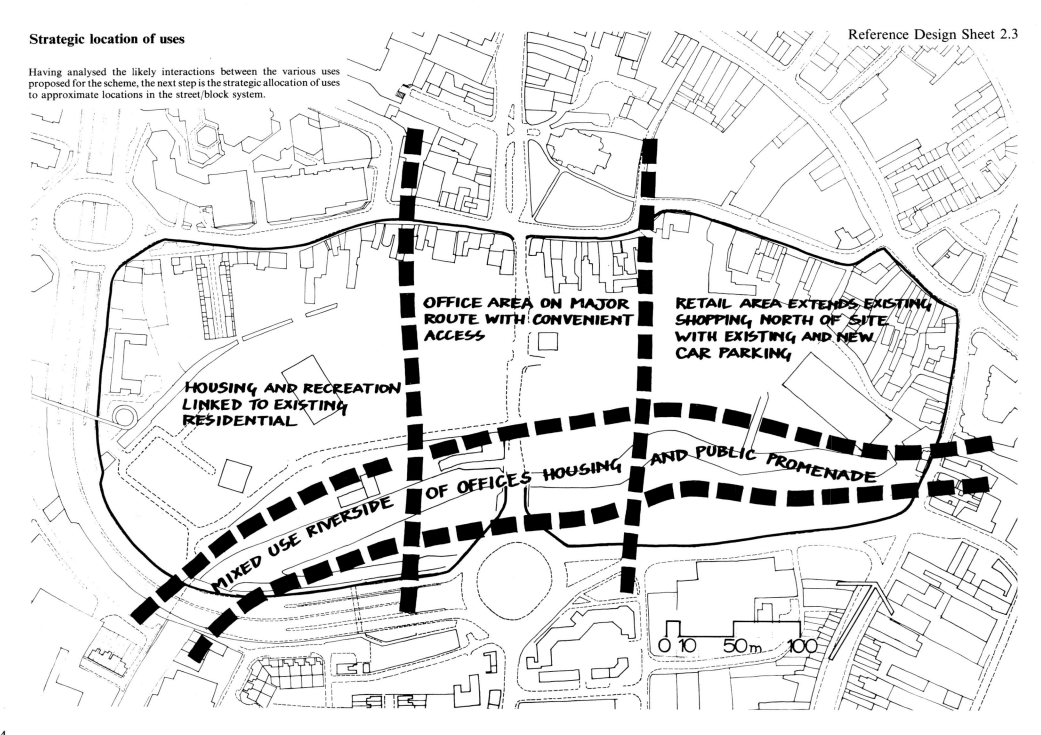

OFFICE AREA ON MAJOR ROUTE WITH CONVENIENT ACCESS

RETAIL AREA EXTENDS EXISTING SHOPPING NORTH OF SITE WITH EXISTING AND NEW CAR PARKING

HOUSING AND RECREATION LINKED TO EXISTING RESIDENTIAL

MIXED USE RIVERSIDE OF OFFICES HOUSING AND PUBLIC PROMENADE

0 10 50 m 100

The shopping layout

Once the various uses are roughly located within the street/block system, it is time to pay particular attention to the positioning of magnets to support any secondary uses which need pedestrian flows.

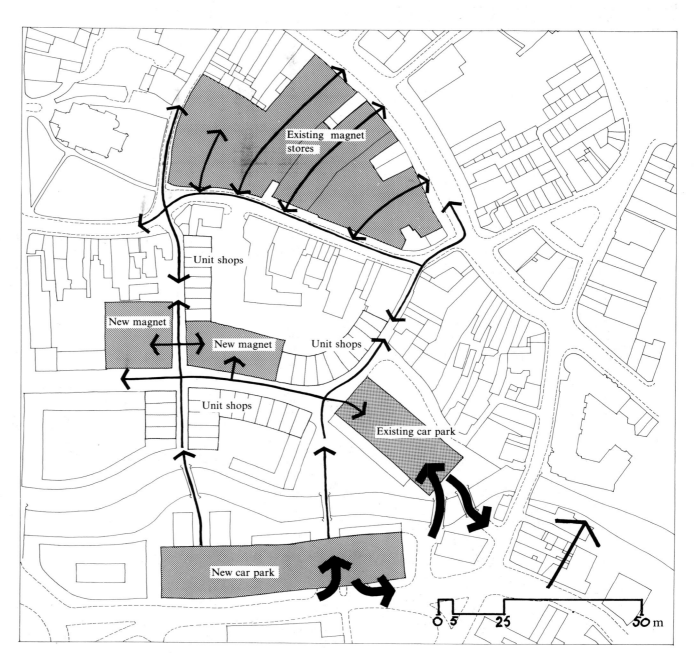

115

Feasibility check

Having established a revised layout to support variety, the next step is to check whether or not the proposal is financially viable.

The data for calculating the costs and values of the various elements in the scheme were taken from builders' price books[1] and from estate agents, as discussed in Design Sheets 2.4 and 2.5. These data, together with the formulae for calculating total costs and values, were then put into a 'Visicalc' electronic spreadsheet, using an Apple 2-plus micro-computer[2]. This enabled the economic feasibility of the scheme to be continually monitored as the design developed. A print-out from the Visicalc is shown below.

	NUMBER	GROSS AREA	NETT AREA	SALE PRICE	RENT/ SQ. M.	RENT ROLL	YIELD%	VALUE	UNIT COST	CONST COST	LAND COST	FEES PERCENT	TOTAL FEES	FUND%	CONTRACT PERIOD	VOIDS PERIOD	INTEREST CONST	INTEREST VOIDS	TOTAL COST	DEV'S PROFIT	PROFIT%
SHOPS A		770	770		340	261800	6.25	4188800	250	192500		12	23100								
SHOPS B		770	770		170	130900	6.25	2094400	250	192500		12	23100								
SHOPS C		770	770		85	65450	6.25	1047200	250	192500		12	23100								
F F STORE		900	900		42.5	38250	6.25	612000	250	225000		12	27000								
SUPERMARK		2100	2100		120	252000	6.25	4032000	300	630000		12	75600								
STORE		3268	3268		120	392160	7.5	5228800	350	1143800		12	137256								
OFFICES		21500	17200		130	2236000	6.25	35776000	500	10750000		12	1290000								
FLATS	100	6000	5000	30000				3000000	330	1980000		12	237600								
HOUSES	30	2400	2400	50000				1500000	320	768000		12	92160								
LEISURE		2520	2016		0				440	1108800		12	133056								
MALT HO		1320			0				500	660000		12	79200								
MSCARPARK		18816			0				120	2257920											
MARINA					0					900000		12	108000								
TOTALS		61134						57479200		21001020	11000000		2249172	12	4	.5	11043579	2717626.	48011397	9467803.	19.71991

Street classification and dimensions

With the exception of Bridge Street, all the streets in the scheme are classified as access roads. Because of the different uses to which they give access, these are further classified into *all-purpose* and *residential* access roads.

Carriageway widths

Each access road carries a different intensity of traffic. This must be calculated, as shown in Figure 1, so that carriageway widths can be determined.

Junction spacings

Most of the junctions in the proposed layout are already satisfactorily spaced for the street types concerned: only those onto Bridge Street are open to question. However, our own traffic consultants suggest that the use of mini-roundabouts - already proposed, by the highway authority, for the existing Bridge Street junctions - should make the new junctions feasible. This will need detailed negotiation with the highway authority as the design develops.

Detailed junction design

Allowing for the traffic-slowing effects of the mini-roundabouts on Bridge Street, all the necessary set-backs, visibility splays and junction radii can be accommodated in the design of the footpaths concerned. There is no need to alter the building lines already proposed.

The pedestrian network

Nearly all the streets are designed for combined vehicular and pedestrian use. They will later be designed in detail so that vehicles will not dominate pedestrian users. Because of the limitations of some of the existing accesses to the site, and the positions of existing buildings to be retained, some pedestrian-only routes are unavoidably proposed. All of them, however, are treated as truly public spaces, defined by building fronts.

In addition, the new shopping routes are restricted to pedestrian use to meet the developer's requirements. These are nonetheless wide enough to become combined pedestrian/vehicle spaces should the opportunity later arise.

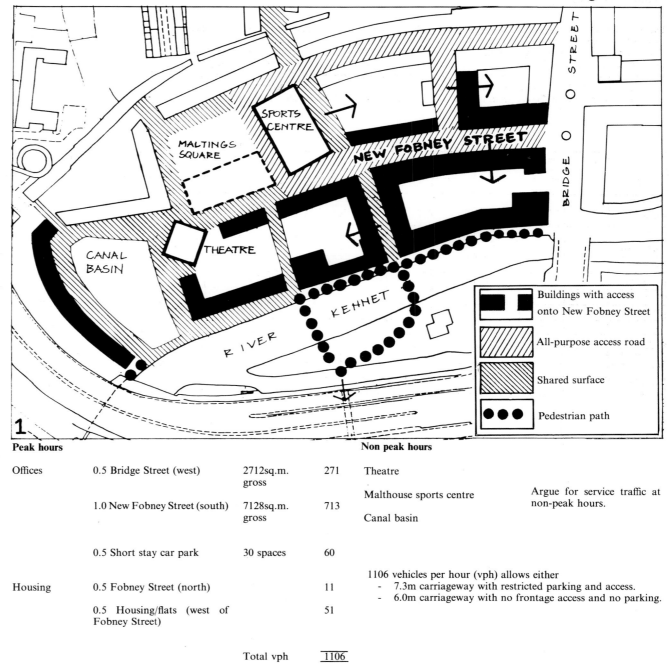

1

Legend:
- Buildings with access onto New Fobney Street
- All-purpose access road
- Shared surface
- Pedestrian path

Peak hours				Non peak hours
Offices	0.5 Bridge Street (west)	2712 sq.m. gross	271	Theatre
	1.0 New Fobney Street (south)	7128 sq.m. gross	713	Malthouse sports centre
				Canal basin
	0.5 Short stay car park	30 spaces	60	
Housing	0.5 Fobney Street (north)		11	
	0.5 Housing/flats (west of Fobney Street)		51	
	Total vph		1106	

Argue for service traffic at non-peak hours.

1106 vehicles per hour (vph) allows either
- 7.3m carriageway with restricted parking and access.
- 6.0m carriageway with no frontage access and no parking.

Block size check

Now that proposals for use-locations and street layout are firming up, it is necessary to check that the various blocks are of an adequate size to house the uses proposed for them.

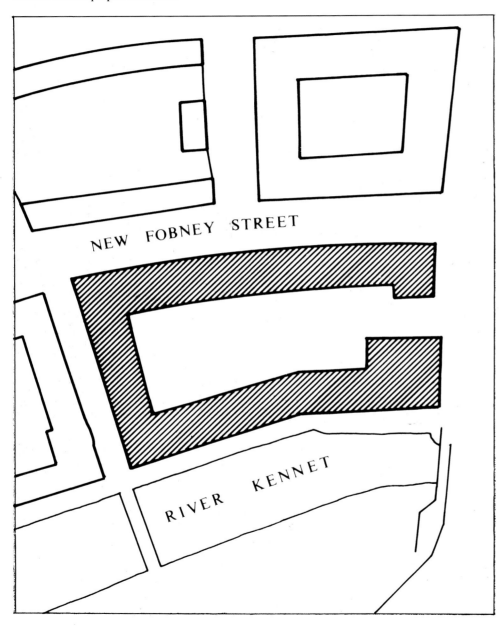

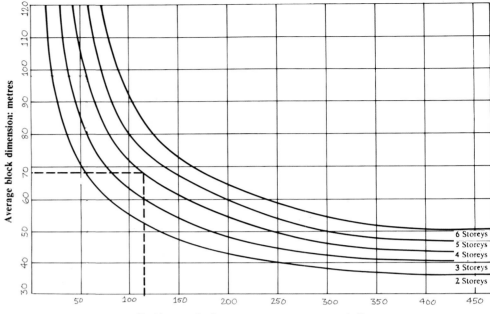

Parking standard: cars per gross square metre built space

This block contains only offices. It is 4 storeys high.

Total gross floorspace (approx) = 7890 sq.m

Parking standard required: 1 car per 100 sq.m. nett floorspace; of which 50% will be in multi-storey car park.

On site, therefore, we need 1 car per 200 sq.m. nett floorspace.

Average block dimension = (90 + 46) ÷ 2 = 68 m. From graph above (see Design Sheet 1.4) parking standard available per car = 120 sq.m.

This is more than adequate, and allows extra space for planting, seating and so forth.

Summary: the scheme after considering permeability and variety

By this stage, the layout of streets and blocks has been decided in some detail, and the volumes of the buildings housing the various uses have been checked, to produce a financially feasible scheme. This plan summarises the design decisions made so far.

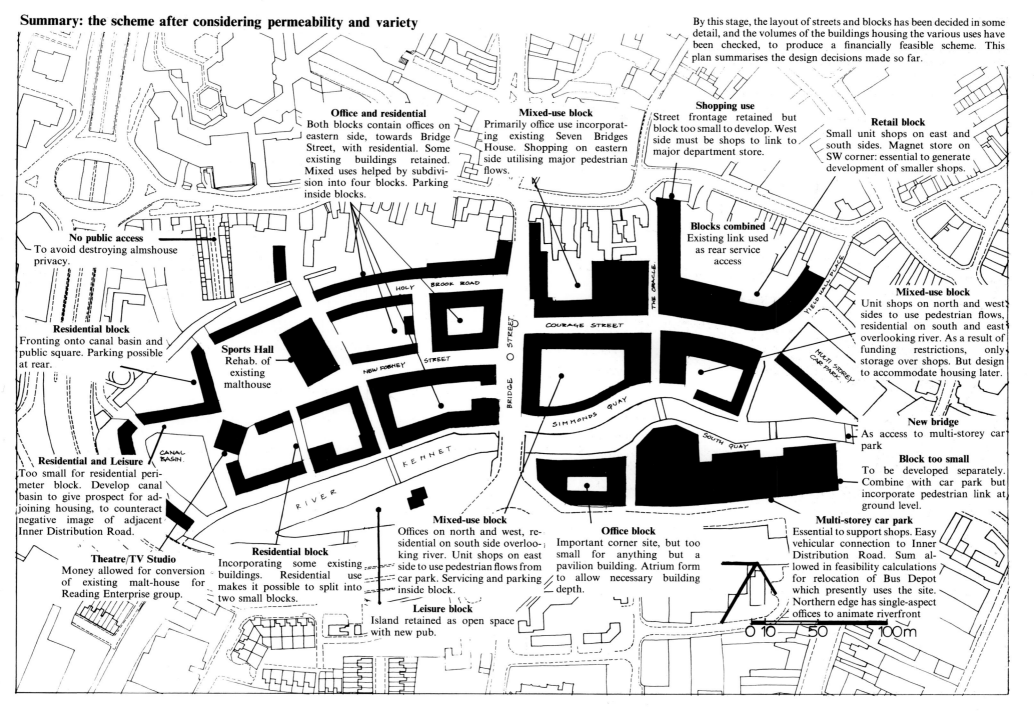

Office and residential
Both blocks contain offices on eastern side, towards Bridge Street, with residential. Some existing buildings retained. Mixed uses helped by subdivision into four blocks. Parking inside blocks.

Mixed-use block
Primarily office use incorporating existing Seven Bridges House. Shopping on eastern side utilising major pedestrian flows.

Shopping use
Street frontage retained but block too small to develop. West side must be shops to link to major department store.

Retail block
Small unit shops on east and south sides. Magnet store on SW corner: essential to generate development of smaller shops.

No public access
To avoid destroying almshouse privacy.

Blocks combined
Existing link used as rear service access

Mixed-use block
Unit shops on north and west sides to use pedestrian flows, residential on south and east overlooking river. As a result of funding restrictions, only storage over shops. But design to accommodate housing later.

Residential block
Fronting onto canal basin and public square. Parking possible at rear.

Sports Hall
Rehab. of existing malthouse

Residential and Leisure
Too small for residential perimeter block. Develop canal basin to give prospect for adjoining housing, to counteract negative image of adjacent Inner Distribution Road.

New bridge
As access to multi-storey car park

Block too small
To be developed separately. Combine with car park but incorporate pedestrian link at ground level.

Theatre/TV Studio
Money allowed for conversion of existing malt-house for Reading Enterprise group.

Residential block
Incorporating some existing buildings. Residential use makes it possible to split into two small blocks.

Mixed-use block
Offices on north and west, residential on south side overlooking river. Unit shops on east side to use pedestrian flows from car park. Servicing and parking inside block.

Office block
Important corner site, but too small for anything but a pavilion building. Atrium form to allow necessary building depth.

Multi-storey car park
Essential to support shops. Easy vehicular connection to Inner Distribution Road. Sum allowed in feasibility calculations for relocation of Bus Depot which presently uses the site. Northern edge has single-aspect offices to animate riverfront

Leisure block
Island retained as open space with new pub.

0 10   50   100m

119

Legibility analysis

Now that the design has been developed to support variety, the next step is to consider the legibility of the scheme. As a first step, we analysed the legibility of the site and its surroundings as they exist at present.

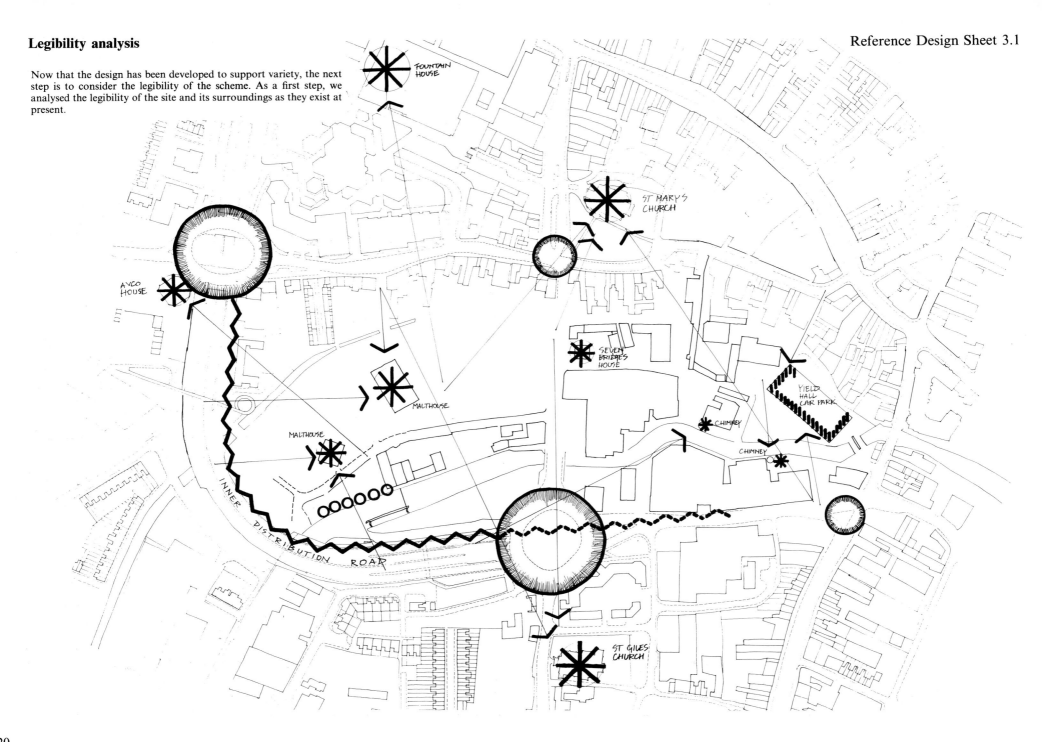

The next step was to check our own initial analysis of legibility against the views of other people. The results of the process are tabulated in Figure 1. This shows that the designer's own analysis predicted 8 out of the 10 most frequently-mentioned features; but the checking process greatly enriched the designers' eventual view of the site's legibility potential, as summarised on page 125.

RESPONDENT	A	B	C	D	E	F	G	H	J	K	L		
GUN STREET	✓	✓	✓	✓	✓	✓	✓	✓	✓	✓	✓	11	
BRIDGE STREET	✓	✓	✓	✓	✓	✓	✓	✓	✓	✓	✓	11	✳
RIVER KENNET	✓	✓	✓	✓	✓	✓	✓	✓	✓	✓	✓	11	✳
ROUNDABOUT	✓	✓	✓	✓	✓	✓	✓		✓	✓	✓	10	✳
COURAGE'S OFFICES	✓	✓	✓	✓	✓	✓	✓		✓	✓	✓	10	
BUS DEPOT	✓	✓	·		✓	✓	✓	✓	✓	✓	✓	10	✳
INNER DISTRIBUTION ROAD		✓	✓	✓	✓	✓	✓	✓	✓		✓	9	✳
SEVEN BRIDGES HOUSE		✓	✓	✓	✓	✓	✓	✓	✓	✓		9	✳
CASTLE STREET	✓	✓	✓		✓	✓	✓	✓	✓			9	
SMALL MALTHOUSE	✓	✓		✓	✓		✓		✓	✓		7	✳
FOBNEY STREET	✓	✓		✓	✓		✓			✓		7	
CASTLE STREET / IDR JUNCTION	✓			✓		✓	✓		✓		✓	6	✳
GEORGIAN: CASTLE ST/GUN ST		✓			✓	✓	✓		✓	✓		6	
THE ORACLE	✓	✓	✓					✓	✓	✓		6	
BRIDGE ST: CORNER BUILDING		✓	✓			✓	✓		✓	✓		6	
ST MARYS BUTTS	✓		✓			✓	✓			✓	✓	6	✳
BREWERY DISTRICT		✓		✓			✓	✓	✓		✓	6	✳
GAS LANE	✓	✓		✓		✓				✓		5	
HOLY BROOK		✓	✓			✓	✓	✓				5	
'VACANT SITE' DISTRICT		✓		✓	✓				✓	✓		5	
MOBILE HOMES	✓			✓	✓	✓					✓	5	
SALVATION ARMY HOSTEL	✓			✓	✓				✓			5	
RIVERSIDE GREEN SPACE				✓	✓	✓					✓	4	
WHITBREAD COURT	✓				✓	✓					✓	4	
MULTI-STOREY CAR PARK	✓				✓		✓			✓		4	✳
ALMSHOUSES		✓		✓			✓		✓			4	
ST GILES' CHURCH	✓	✓				✓				✓		4	✳
LARGE MALTHOUSE		✓	✓				✓					3	✳
BREWERY BRIDGE (DEMOLISHED)	✓	✓								✓		3	
YIELD HALL LANE	✓						✓			✓		3	
CASTLE ST: BREWERY ENTRANCE	✓			✓				✓				3	
HEELAS/DEBENHAMS/B.H.S.	✓					✓	✓					3	
LOCK ON RIVER KENNET						✓	✓	✓				3	
CIVIC CENTRE						✓	✓					2	
IDR PEDESTRIAN BRIDGE				✓				✓				2	
BRIDGE ST PUB (DEMOLISHED)				✓				✓				2	
GEORGE HOTEL	✓									✓		2	
SUBWAY				✓							✓	2	
MAP ANALYSIS	9 ETC = NO OP MAPS WITH THIS ELEMENT												
	✳ = ELEMENT NOTED BY DESIGNER												

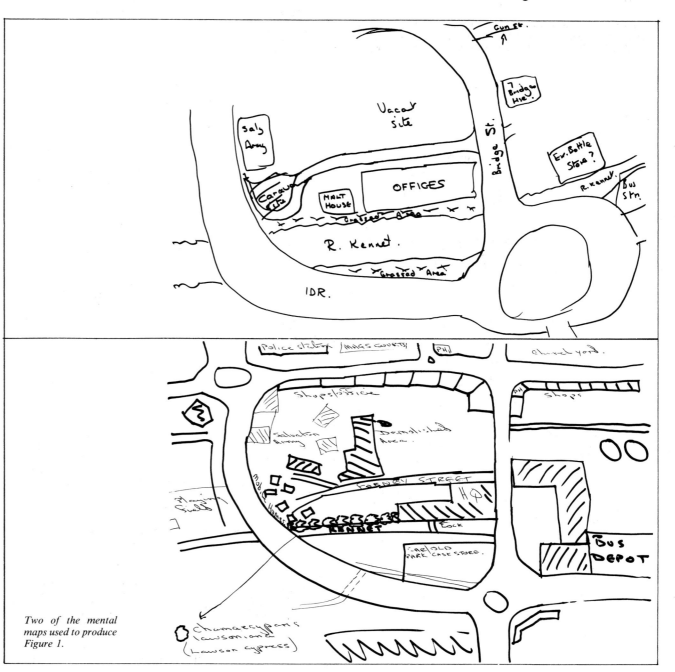

Two of the mental maps used to produce Figure 1.

District implications

The western part of the site is cut off from its eastern and southern surroundings, and is different in use from the district to the north of it: it was decided to treat this as a new district in its own right.

In order to encourage the shops to take off in an area not previously associated with shopping, the developers wanted the eastern part of the site to read as part of the established commercial district.

Figure 1 summarises these district decisions.

The central commercial district has strong path themes: the north - south streets have very much smaller dimensions than those running east - west. The forms of the streets were analysed, as shown in Figure 2, to arrive at a vocabulary of street forms to be used in the eastern part of the site.

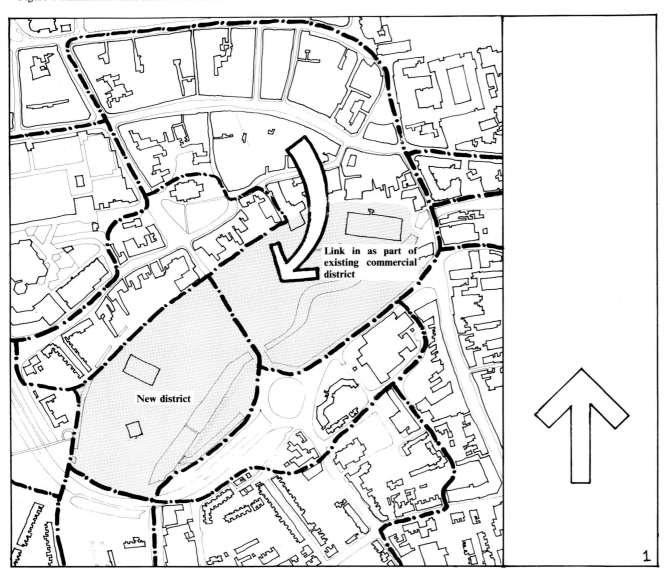

Link in as part of existing commercial district

New district

1

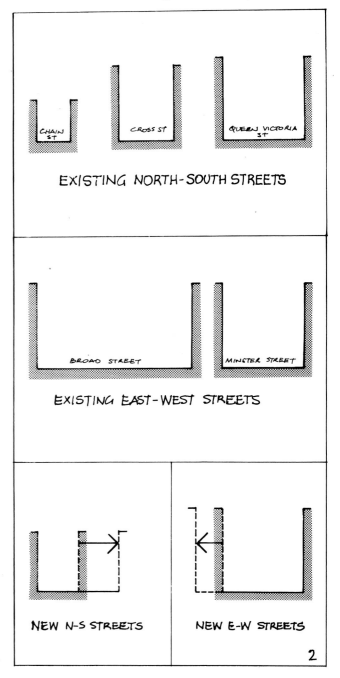

2

Nodes and marker sequences

Figures 1, 2 and 3 illustrate legibility decisions related to different types of nodes.

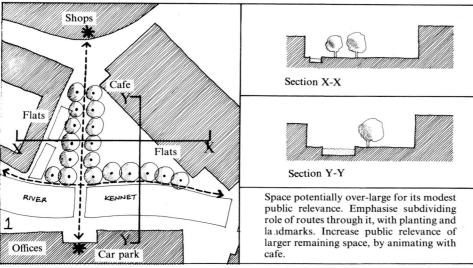

Section X-X

Section Y-Y

Space potentially over-large for its modest public relevance. Emphasise subdividing role of routes through it, with planting and landmarks. Increase public relevance of larger remaining space, by animating with cafe.

The Triangle

A space whose large size - necessary because of the existing car park and minor waterway - is inappropriate, in legibility terms, for its modest functional significance.

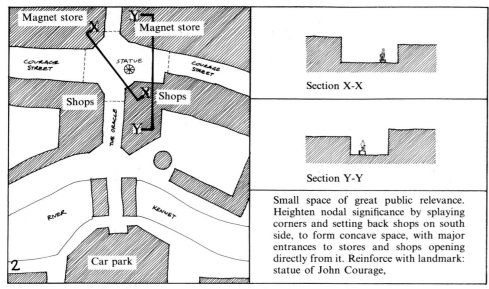

Section X-X

Section Y-Y

Small space of great public relevance. Heighten nodal significance by splaying corners and setting back shops on south side, to form concave space, with major entrances to stores and shops opening directly from it. Reinforce with landmark: statue of John Courage,

The Oracle / Courage Street intersection

A space of key significance to shoppers.

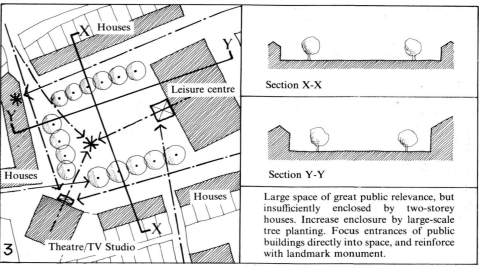

Section X-X

Section Y-Y

Large space of great public relevance, but insufficiently enclosed by two-storey houses. Increase enclosure by large-scale tree planting. Focus entrances of public buildings directly into space, and reinforce with landmark monument.

Maltings Place

A large space, appropriate in legibility terms for its high public relevance, but weakly enclosed by two-storey housing on two sides.

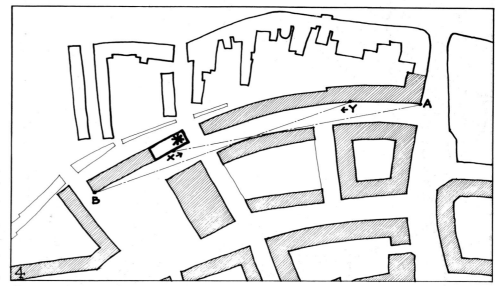

Courage street / Yield Hall Lane

Because of its curved planform, the Courage Street / Yield Hall Lane sequence needs an additional marker. Figure 4 shows how this was positioned.

Summary: layout adjusted to achieve legibility

By now, the design has been developed to make it as legible as possible.
The revised proposals are summarised in this axonometric.

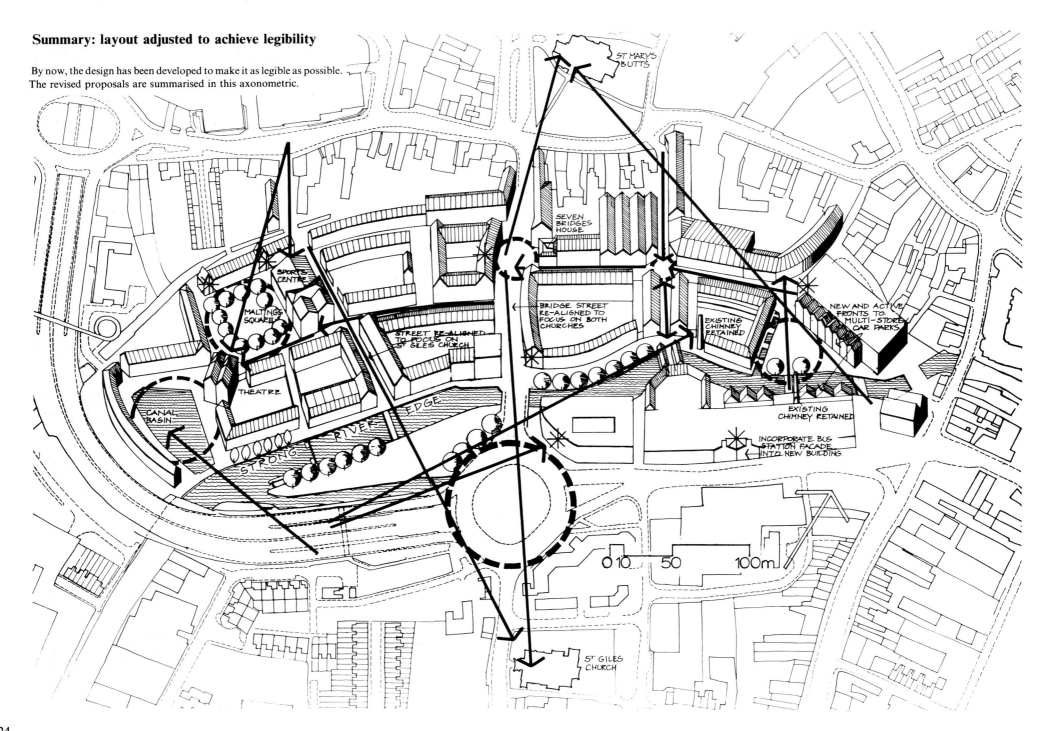

ST MARY'S BUTTS

SEVEN BRIDGES HOUSE

SPORTS CENTRE

MALTINGS SQUARE

STREET RE-ALIGNED TO FOCUS ON ST GILES CHURCH

BRIDGE STREET RE-ALIGNED TO FOCUS ON BOTH CHURCHES

EXISTING CHIMNEY RETAINED

NEW AND ACTIVE FRONTS TO MULTI-STOREY CAR PARKS

THEATRE

CANAL BASIN

RIVER EDGE

STRONG

EXISTING CHIMNEY RETAINED

INCORPORATE BUS STATION FACADE INTO NEW BUILDING

0 10   50   100m

ST GILES CHURCH

124

Layout adjusted to achieve legibility: model

A working model, quickly and cheaply made from scrap polystyrene, allows the emerging design to be evaluated from a variety of viewing positions.

View from the south west: new retail and office areas with riverfront housing.

View from the city centre, along Bridge Street, towards the River Kennet.

General view from the north east.

Riverfront housing, Maltings Place and the new canal basin.

Robust terraced housing

This sheet shows the decisions made to support robustness, in the design of the terraced houses and gardens on the south side of Maltings Place.

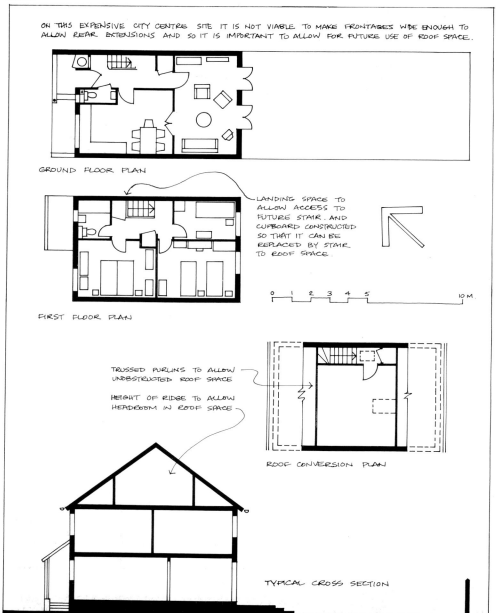

ON THIS EXPENSIVE CITY CENTRE SITE IT IS NOT VIABLE TO MAKE FRONTAGES WIDE ENOUGH TO ALLOW REAR EXTENSIONS AND SO IT IS IMPORTANT TO ALLOW FOR FUTURE USE OF ROOF SPACE.

GROUND FLOOR PLAN

LANDING SPACE TO ALLOW ACCESS TO FUTURE STAIR. AND CUPBOARD CONSTRUCTED SO THAT IT CAN BE REPLACED BY STAIR TO ROOF SPACE.

FIRST FLOOR PLAN

0 1 2 3 4 5   10 M

TRUSSED PURLINS TO ALLOW UNOBSTRUCTED ROOF SPACE

HEIGHT OF RIDGE TO ALLOW HEADROOM IN ROOF SPACE

ROOF CONVERSION PLAN

TYPICAL CROSS SECTION

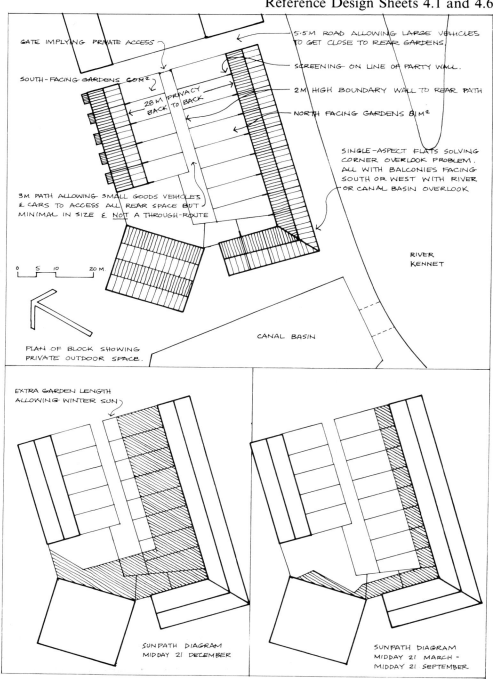

GATE IMPLYING PRIVATE ACCESS

5.5M ROAD ALLOWING LARGE VEHICLES TO GET CLOSE TO REAR GARDENS.

SOUTH-FACING GARDENS 60M²

SCREENING ON LINE OF PARTY WALL.

28M PRIVACY BACK TO BACK

2M HIGH BOUNDARY WALL TO REAR PATH

NORTH FACING GARDENS 81M²

SINGLE-ASPECT FLATS SOLVING CORNER OVERLOOK PROBLEM. ALL WITH BALCONIES FACING SOUTH OR WEST WITH RIVER OR CANAL BASIN OVERLOOK

3M PATH ALLOWING SMALL GOODS VEHICLES & CARS TO ACCESS ALL REAR SPACE BUT MINIMAL IN SIZE & NOT A THROUGH-ROUTE

0 5 10   20 M

RIVER KENNET

CANAL BASIN

PLAN OF BLOCK SHOWING PRIVATE OUTDOOR SPACE.

EXTRA GARDEN LENGTH ALLOWING WINTER SUN

SUNPATH DIAGRAM MIDDAY 21 DECEMBER

SUNPATH DIAGRAM MIDDAY 21 MARCH - MIDDAY 21 SEPTEMBER

A robust block

This sheet shows the preferred building configuration advocated in Design Sheet 4.2, used to design the mixed-use block fronting Bridge Street, to the south of Courage Street.

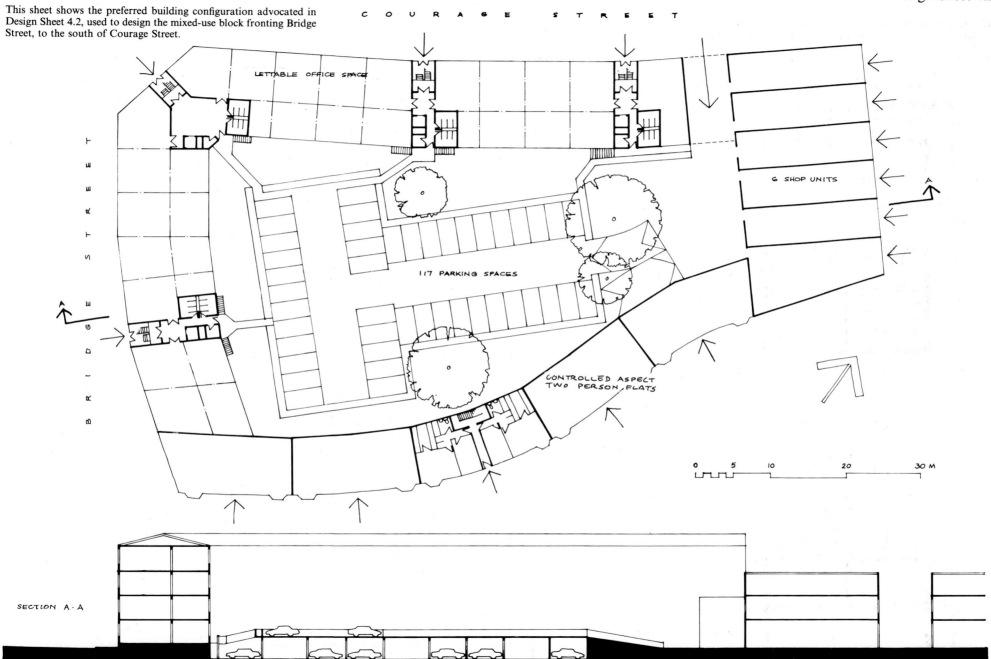

COURAGE STREET

BRIDGE STREET

LETTABLE OFFICE SPACE

117 PARKING SPACES

6 SHOP UNITS

CONTROLLED ASPECT TWO PERSON FLATS

A

0 5 10 20 30 M

SECTION A-A

Robust internal planning

The interior of the Bridge Street offices is now developed further, locating the hard and soft zones to promote large scale robustness.

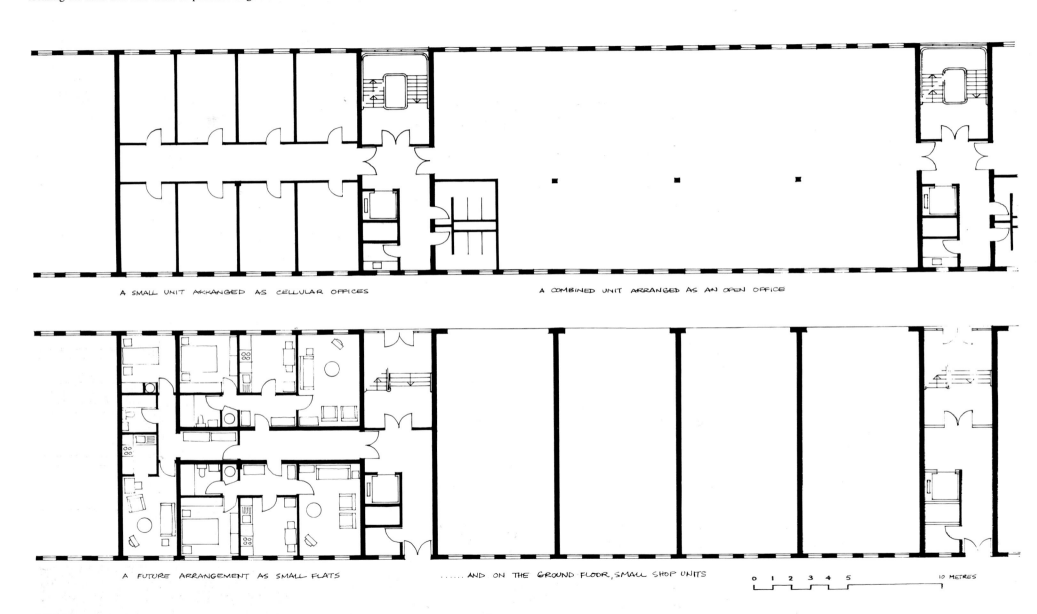

A SMALL UNIT ARRANGED AS CELLULAR OFFICES

A COMBINED UNIT ARRANGED AS AN OPEN OFFICE

A FUTURE ARRANGEMENT AS SMALL FLATS

...... AND ON THE GROUND FLOOR, SMALL SHOP UNITS

0 1 2 3 4 5   10 METRES

Animating the car park edge

The public face of the Yield Hall multi-storey car park is animated by the addition of a single-aspect block of small flats, facing south-west over the river. The interface between the flats and the public space is designed to support a variety of outdoor activities.

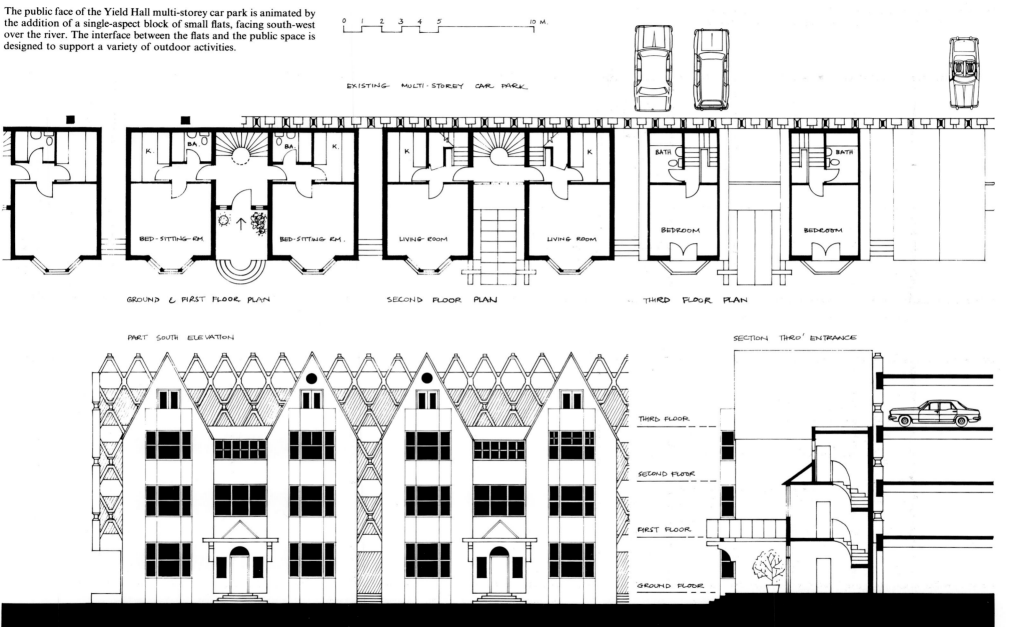

EXISTING MULTI-STOREY CAR PARK

GROUND & FIRST FLOOR PLAN

SECOND FLOOR PLAN

THIRD FLOOR PLAN

PART SOUTH ELEVATION

SECTION THRO' ENTRANCE

THIRD FLOOR

SECOND FLOOR

FIRST FLOOR

GROUND FLOOR

Maltings Place

This sheet shows how the project's largest public outdoor space is designed for a balance between pedestrian and vehicular use, at both public and residential scales.

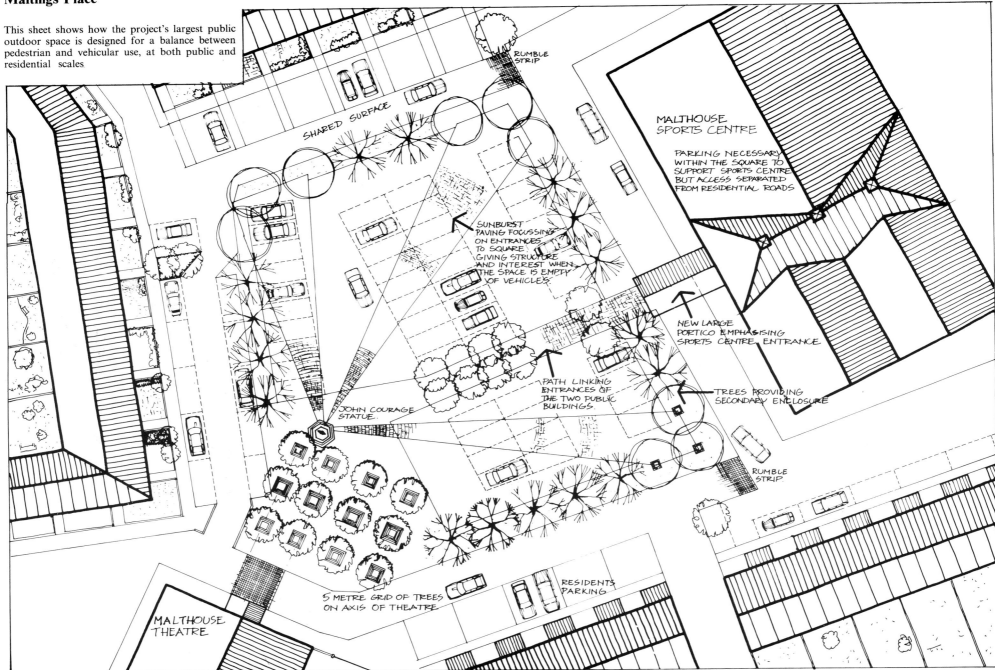

RUMBLE STRIP

SHARED SURFACE

MALTHOUSE SPORTS CENTRE

PARKING NECESSARY WITHIN THE SQUARE TO SUPPORT SPORTS CENTRE BUT ACCESS SEPARATED FROM RESIDENTIAL ROADS

SUNBURST PAVING FOCUSSING ON ENTRANCES TO SQUARE GIVING STRUCTURE AND INTEREST WHEN THE SPACE IS EMPTY OF VEHICLES

NEW LARGE PORTICO EMPHASISING SPORTS CENTRE ENTRANCE

PATH LINKING ENTRANCES OF THE TWO PUBLIC BUILDINGS.

TREES PROVIDING SECONDARY ENCLOSURE

JOHN COURAGE STATUE.

RUMBLE STRIP.

RESIDENTS PARKING

5 METRE GRID OF TREES ON AXIS OF THEATRE

MALTHOUSE THEATRE

Visual appropriateness: a performance specification

This sheet sets out the objectives to be achieved in designing the street elevations of the mixed-use block fronting Bridge Street, south of Courage Street.

Courage Street and Bridge Street elevations.

Objectives to support variety:
- to be interpreted, by the widest possible public, as a part of Reading's established commercial area.
- to be interpreted, by potential investment institutions, as an efficient modern office building.

Objective to support legibility:
- to be interpreted, by the widest possible public, as contrasting with Seven Bridges House.

Objectives to support robustness:
- to be interpreted as appropriate for domestic use, by potential residents of this area of the town.
- to be interpreted as an appropriate base by both large commercial and small professional office tenants.
- the ground floor to be interpreted as an appropriate base by potential office or shop tenants.

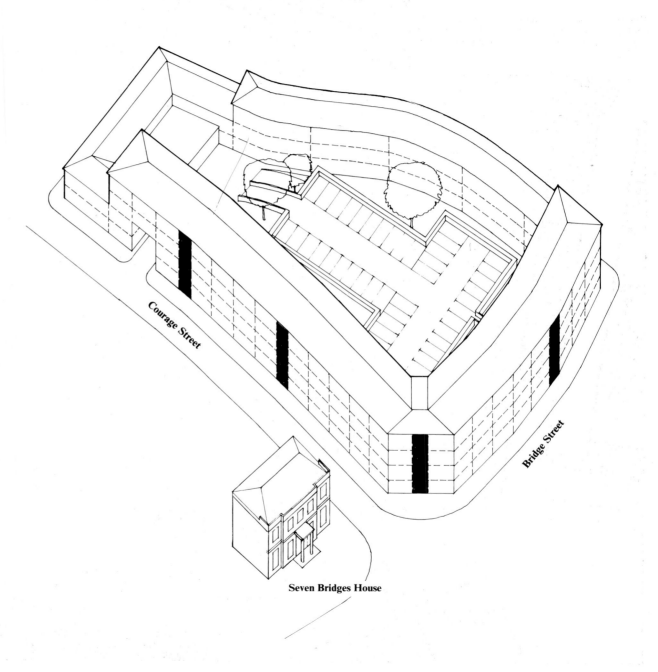

Courage Street

Bridge Street

Seven Bridges House

Using large-scale cues

Here the basis of the Bridge Street office elevation is developed, using large-scale cues to achieve the objectives outlined on the previous page.

Reference Design Sheets 5.2 - 5.6

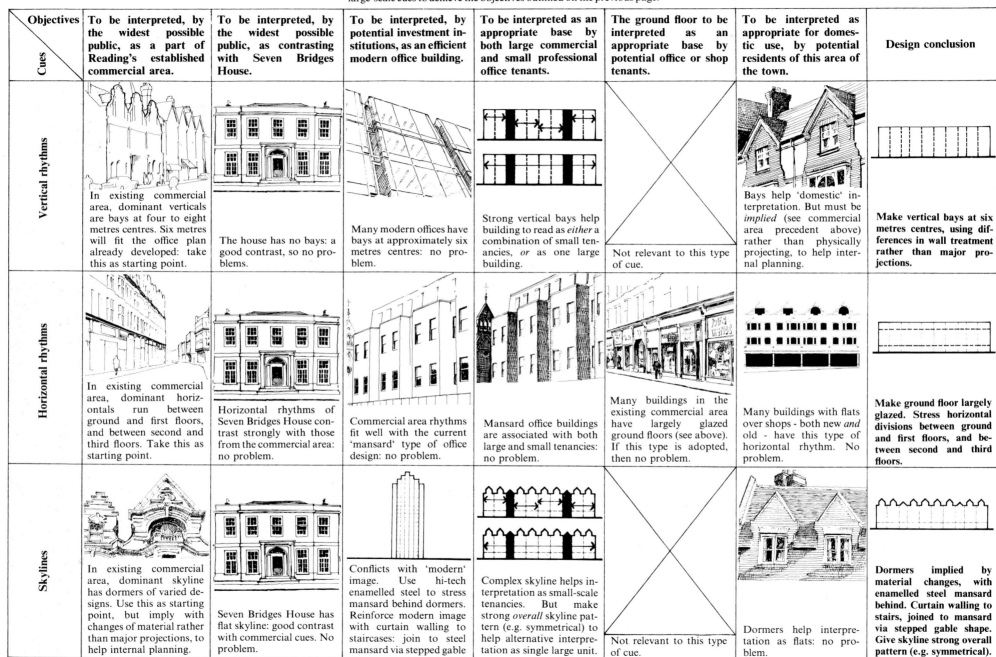

Objectives / Cues	To be interpreted, by the widest possible public, as a part of Reading's established commercial area.	To be interpreted, by the widest possible public, as contrasting with Seven Bridges House.	To be interpreted, by potential investment institutions, as an efficient modern office building.	To be interpreted as an appropriate base by both large commercial and small professional office tenants.	The ground floor to be interpreted as an appropriate base by potential office or shop tenants.	To be interpreted as appropriate for domestic use, by potential residents of this area of the town.	Design conclusion
Vertical rhythms	In existing commercial area, dominant verticals are bays at four to eight metres centres. Six metres will fit the office plan already developed: take this as starting point.	The house has no bays: a good contrast, so no problems.	Many modern offices have bays at approximately six metres centres: no problem.	Strong vertical bays help building to read as *either* a combination of small tenancies, *or* as one large building.	Not relevant to this type of cue.	Bays help 'domestic' interpretation. But must be *implied* (see commercial area precedent above) rather than physically projecting, to help internal planning.	Make vertical bays at six metres centres, using differences in wall treatment rather than major projections.
Horizontal rhythms	In existing commercial area, dominant horizontals run between ground and first floors, and between second and third floors. Take this as starting point.	Horizontal rhythms of Seven Bridges House contrast strongly with those from the commercial area: no problem.	Commercial area rhythms fit well with the current 'mansard' type of office design: no problem.	Mansard office buildings are associated with both large and small tenancies: no problem.	Many buildings in the existing commercial area have largely glazed ground floors (see above). If this type is adopted, then no problem.	Many buildings with flats over shops - both new *and* old - have this type of horizontal rhythm. No problem.	Make ground floor largely glazed. Stress horizontal divisions between ground and first floors, and between second and third floors.
Skylines	In existing commercial area, dominant skyline has dormers of varied designs. Use this as starting point, but imply with changes of material rather than major projections, to help internal planning.	Seven Bridges House has flat skyline: good contrast with commercial cues. No problem.	Conflicts with 'modern' image. Use hi-tech enamelled steel to stress mansard behind dormers. Reinforce modern image with curtain walling to staircases: join to steel mansard via stepped gable	Complex skyline helps interpretation as small-scale tenancies. But make strong *overall* skyline pattern (e.g. symmetrical) to help alternative interpretation as single large unit.	Not relevant to this type of cue.	Dormers help interpretation as flats: no problem.	Dormers implied by material changes, with enamelled steel mansard behind. Curtain walling to stairs, joined to mansard via stepped gable shape. Give skyline strong overall pattern (e.g. symmetrical).

Using small-scale cues

The further build-up of the Bridge Street office elevation, using small-scale cues to finalise the design developed on the previous page.

Cues \ Objectives	To be interpreted, by the widest possible public, as a part of Reading's established commercial area.	To be interpreted, by the widest possible public, as contrasting with Seven Bridges House.	To be interpreted, by potential investment institutions, as an efficient modern office building.	To be interpreted as an appropriate base by both large commercial and small professional office tenants.	The ground floor to be interpreted as an appropriate base by potential office or shop tenants.	To be interpreted as appropriate for domestic use, by potential residents of this area of the town.	Design conclusion
Windows	In existing commercial area, windows are mostly vertical sliders, without intermediate glazing bars, in vertical openings between 1.5:1 and 2.5:1 proportion. Use this as a starting point.	Potentially too similar. To contrast, use tallest commercial area proportion and avoid intermediate glazing bars and timber construction.	Many modern offices use metal vertical sliding sashes: no problem.	This type of window is associated with both large and small tenancies:no problem.	not relevant to this type of cue.	No serious problem with domestic image, but keep windows as small as possible consistent with 'modern office' image: four per bay suits both flat and office plans	Use four metal sliding sashes per bay, without intermediate glazing bars, in vertical openings about 2:1 proportion.
Wall details	Typical walls in existing commercial area are of horizontally-striped brick / stone. Colours red, yellow, white. Use this as starting point.	To contrast with red brick of Seven Bridges House, use yellow brick, or white block or stone, as wall material.	Problem: 'old-fashioned' associations. Could be overcome by emphasising 'modern' curtain walling in key positions: doors, access halls.	'Corporate' tenants can relate to curtain walling, 'professional' users to traditional elements. No problem, but needs care to get balance right.	Not relevant to this type of cue.	Reading has many precedents for housing with striped brick walls. Blockwork likely to be interpreted as 'cheap': use mostly brick.	Make walls of yellow brick, with contrasting bands of brick or (in small quantities) facing block or (economics permitting) light-coloured stone.
Door and ground level details	Existing commercial area has varied entrances: no obvious cues. Ground floor piers mostly square, in polished granite, with contrasting colours at base and head.	Largely glazed ground floor, with granite piers, will contrast strongly with Seven Bridges House. No problem.	To avoid granite piers being interpreted as 'old-fashioned', make infill glazing of obviously 'modern' design.	Vary entrance designs, to give small tenants their own unique front door. But vary within strong overall pattern (e.g. symmetrical) to help alternative interpretation as single large building.	Ground floor must support interpretation as shops or offices. A square grid of glazing has associations with both uses.	Not relevant to this type of cue.	Make square piers in polished granite. Infill with hi-tech glazing, on a square grid. Vary entrance designs, within strong overall pattern (e.g. symmetrical).

133

Bridge Street offices: the final elevation

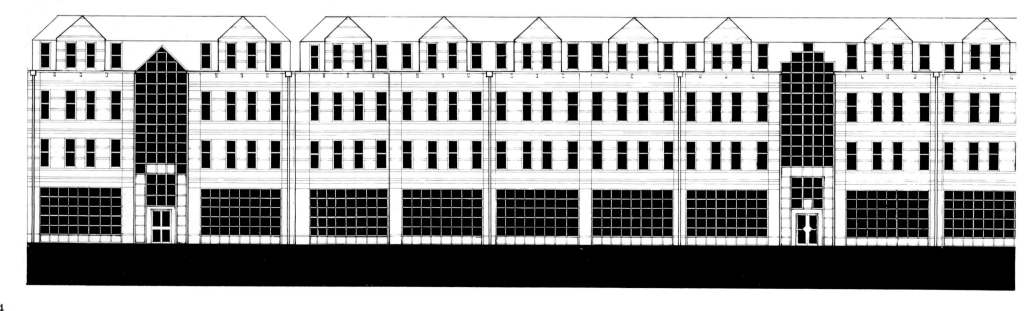

Non-visual richness

This sheet illustrates two places in the scheme which have particular potential for non-visual richness.

Analysing viewing positions

This sheet takes the Bridge Street elevation developed in the previous pages, and analyses the various positions from which it can be seen.

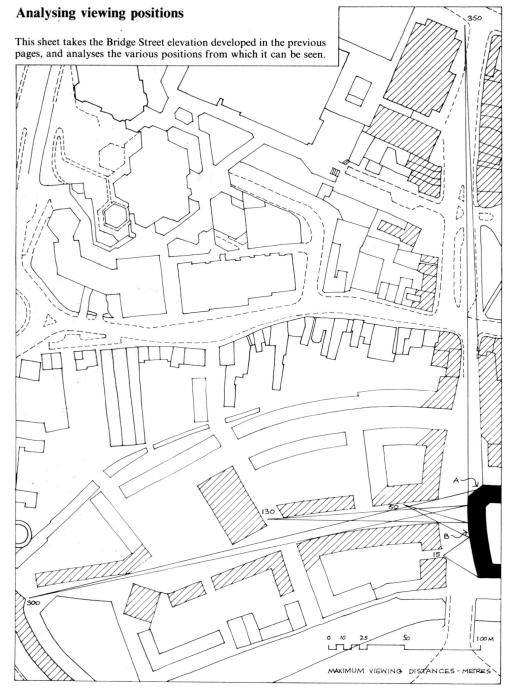

MAXIMUM VIEWING DISTANCES - METRES

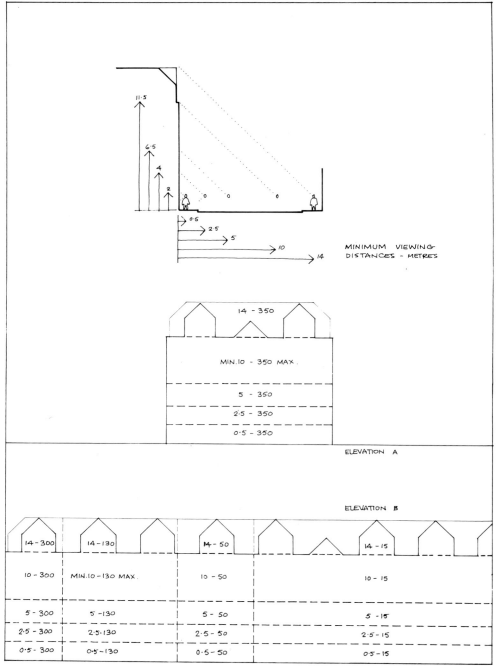

MINIMUM VIEWING DISTANCES - METRES

ELEVATION A

ELEVATION B

The elevation at various distances

Having worked out the viewing positions for the Bridge Street elevation, the next step is to make drawings to simulate what could be seen from the various distances concerned.

THE DRAWING BELOW AT 1:20, WHEN VIEWED FROM 0.5 M, REPRESENTS PART OF THE GROUND FLOOR VIEWED FROM 10M, I.E. THE VIEW FROM ACROSS BRIDGE STREET. WE HAVE DECIDED NOT TO ADD ANY FURTHER VISUAL EVENTS BECAUSE TO DO SO WOULD CONFLICT WITH OUR EARLIER OBJECTIVES.

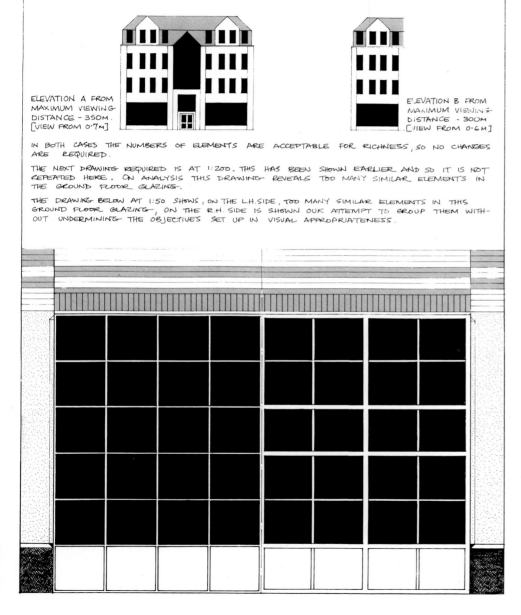

ELEVATION A FROM MAXIMUM VIEWING DISTANCE - 350M. [VIEW FROM 0.7M]

ELEVATION B FROM MAXIMUM VIEWING DISTANCE - 300M. [VIEW FROM 0.6M]

IN BOTH CASES THE NUMBERS OF ELEMENTS ARE ACCEPTABLE FOR RICHNESS, SO NO CHANGES ARE REQUIRED.

THE NEXT DRAWING REQUIRED IS AT 1:200. THIS HAS BEEN SHOWN EARLIER AND SO IT IS NOT REPEATED HERE. ON ANALYSIS THIS DRAWING REVEALS TOO MANY SIMILAR ELEMENTS IN THE GROUND FLOOR GLAZING.

THE DRAWING BELOW AT 1:50 SHOWS, ON THE L.H. SIDE, TOO MANY SIMILAR ELEMENTS IN THIS GROUND FLOOR GLAZING, ON THE R.H. SIDE IS SHOWN OUR ATTEMPT TO GROUP THEM WITHOUT UNDERMINING THE OBJECTIVES SET UP IN VISUAL APPROPRIATENESS.

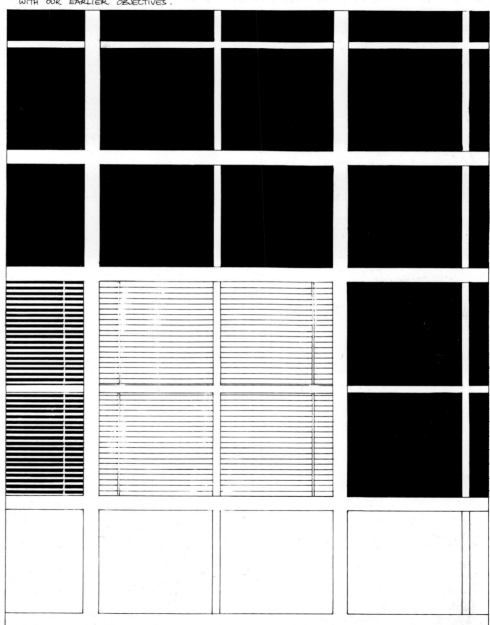

An office entrance

This sheet further develops the richness of one of the Bridge Street office entrances: a position where the building is likely to be seen at close range for considerable periods.

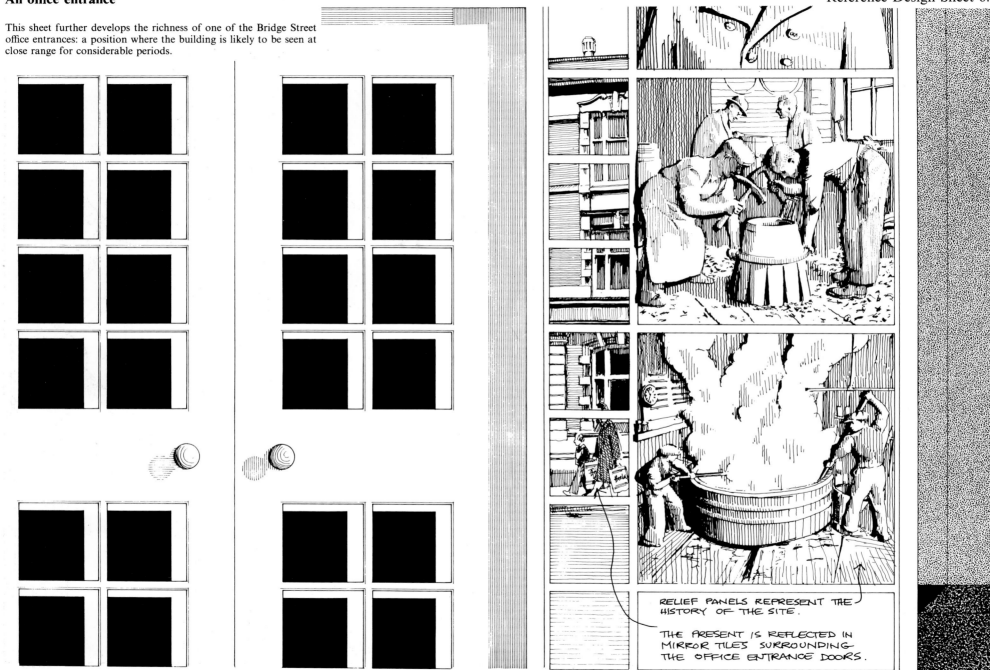

RELIEF PANELS REPRESENT THE HISTORY OF THE SITE.

THE PRESENT IS REFLECTED IN MIRROR TILES SURROUNDING THE OFFICE ENTRANCE DOORS.

Design decisions to support the interior personalisation of the Maltings Place houses

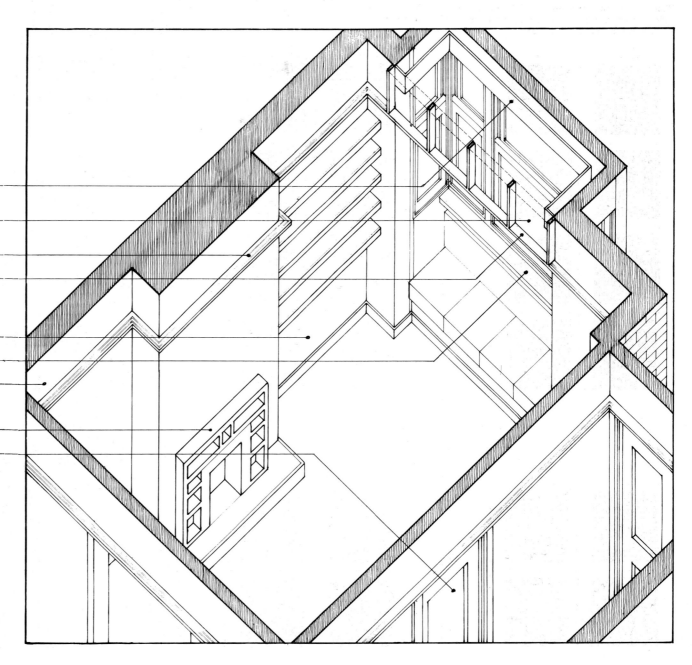

Space above window for pelmet.

Sliding sash windows to allow window boxes. Timber bay framing for easy fixing of window display elements.

Picture rails wide enough to be used as display shelves, with grooved tops to support plates and similar objects.

High-level shelves, particularly good for display of glass objects seen against the light. Also helps to define window nook as alternative focus to the room (see Design Sheet 4.5)

Alcove in brick walls convenient to locate and fix shelves.

Wide tiled cill for display.

Zone between ceiling and wall, which can be painted in with either, to change the apparent proportions of the room.

Fireplace with shelves and niches for display.

Panelled timber doors, which can be 'picked out' in different ways.

This sheet shows the design decisions made to support the exterior personalisation potential of houses along the north side of Maltings Place.

Space around door sufficient to take lamps, nameplates or future porches. Existing porch has ledge over bin store, for display of plants.

Walls which might have climbing plants are of brick: low maintenance, therefore little interference with growing plants.

Sliding sash windows, and projected first floor external cills, make it easy to install window boxes. Articulated timber bays can be 'picked out' in different ways when they are painted.

Horizontal brick bands are interrupted by bays and downpipes. If the bands are obscured by new wall finishes, this will have a minimal effect on the image of the terrace as a whole.

Courage Street, looking towards The Oracle

The Triangle, looking towards Courage Street

Notes

Introduction

1. This attitude is deeply ingrained in the tradition of modern architecture. It goes back at least as far as Gottfried Semper, and comes through very strongly in the ideas of Otto Wagner. For discussion of this aspect of Wagner's work, see Geretsegger and Peintner (1979).

Chapter 1.

1. See Dept. of the Environment and Dept. of Transport's Design Bulletin 32 (1977), Dept. of Transport's *Roads in urban areas* (1966), and Noble (1983). Unless otherwise stated, the tables in Design Sheet 1.3 are taken from the above publications. A key point to remember is that these standards are advisory rather than mandatory. They are currently (1985) under review, and research has raised a number of questions about them.

Chapter 2.

1. The many factors affecting the unequal opportunities for car ownership in Great Britain, for example, are explored in Bates (1978, 1981).
2. For further discussion of this topic, see Bentley (1983a).
3. For useful information on cost control, see Bathurst and Butler (1980). For the importance of time considerations, see Heery (1975).
4. A non-technical account of the relationship between building age and condition, rents and patterns of use is given in Jacobs (1965) Chapter 7.
5. See Markus (1979).
6. See, for example, Bowyer (1979).
7. For a discussion of this topic, see Barrett (1979).
8. For a comprehensive list of land uses and building types, see RIBA (1969).
9. See Tutt and Adler (1979) Chapter 17.
10. For a comparison of different valuation approaches, and much other useful information, see Williams (1976).
11. The yield of an investment is simply the relationship expressed as a percentage, that the income derived from it bears to the capital price paid out for it. (Booth, 1984, 34).

12. See, for example, Spon (1984).
13. Bowyer, op. cit.
14. Because of Britain's uncertain economy, it is now the usual practice to make an additional allowance for the cost of short-term finance for a notional period - often taken as six months - during which the project is completed but as yet unlet. This extra cost of interest during the letting period is calculated as 100% of the total project cost, for six months, at the rate of interest used for the rest of the cost calculations.
    For example:
    Interest during letting period:
    100% of £10,000,000 for 6 months at 12.75% = £637,500.
15. See Baker (1976). For information on the provisions of the Housing Act, 1980, which apply to registered Housing Associations, see Housing Corporation Circular 11/80 (Sept 1980).
16. The TIC system was introduced in 1982. TIC matrices for both rehab and new build are revised quarterly, and published by the Housing Corporation. For detailed guidance on preparing schemes for housing association development, see the current Housing Corporation Schemework Procedure Guide, which is frequently updated.

Chapter 3.

1. For a discussion of the value of achieving congruence between patterns of form and activity, see Steinitz (1968).
2. Lynch (1975) p 21, Fig 5.
3. Ibid, chapter 3.
4. Ibid, p169. Fig 56.
5. Ibid, pp 144 et seq.

Chapter 4.

1. See, for example, Tutt and Adler (1979) or Neufert (1970).
2. The importance of tradition and precedent in coping with complex problems is explored in Shils (1981). A similar argument, related specifically to design problems, is put forward by Alexander (1971).
3. Duffy et. al. (1980).
4. Cowan (1963).
5. For a discussion of this topic, see Whyte (1980).
6. Duffy et. al., op. cit.

7. Ibid.
8. A distance based on British Standards and Codes of Practice; allowing a maximum direct escape distance, in one direction only, of 12 metres.
9. Cowan, op. cit.
10. See McLaughlin and McLaughlin (1978), who advise that the average family is reckoned to get a year's vegatables off a plot measuring 9 X 18 metres. This will vary according to climate, soil, expertise and eating requirements, but is useful as a general rule of thumb.
11. The concepts of prospect-refuge theory are discussed in Appleton (1975).
12. See the discussion in Project for Public Spaces (1982).
13. See Grotenhuis (1978) and Royal Dutch Touring Club (1980).
14. Broadly speaking, wind begins to be annoying when its speed exceeds 5 metres per second. The frequency with which this speed is exceeded is used as a simple criterion in making design decisions. See Penwarden and Wise (1975).

Chapter 5.

1. For discussions of this important factor, see Bonta (1979) and Broadbent (1977)
2. See Bernstein (1971), Bourdieu (1980) and Clarke (1973).
3. This is explored in Douglas (1978, 1982).
4. See, for example, Berger and Luckmann (1966) and Moore (1983).
5. For an interesting example, see Hanson and Hillier (1982).
6. See Appleyard (1969) and Moore (op. cit.)

Chapter 6.

1. See, for example, Gibson (1966).
2. This concept is discussed in Miller (1956).
3. For an account of these aspects of the design see Filler (1978).
4. In Greenacre Park and Paley Park - two tiny spaces fronting busy vehicular streets in New York City - falling water is used to create white noise, countering traffic noise intrusion. On occasions when the water has been turned off, people have quickly begun to leave. Greenacre Park is discussed in Project for Public Spaces (1982). See also Whyte, op. cit.

5. For interesting accounts of Halprin's work, see Halprin (1969) and Process Architecture (1978).
6. For a discussion of the characteristics and cultivation of scented plants, see Garland (1984).
7. The Pompidou Centre (sometimes called Centre Beaubourg) in Paris, France was designed by Piano and Rogers, and built in 1977. It is a national cultural centre containing a museum of modern art, a reference library, a centre for industrial design and a centre for music and acoustic research. A public escalator rises along that side of the building which encloses a new public square. See Architectural Review Vol CLXI No. 693, May 1977, pp. 270-294.

Chapter 7.
1. For a discussion of this problem, see Bentley (1983b) pp 8-11.
2. See, for example, Grimsby Borough Council (1976) (reviewed in Building Design, 20.2.76) and Wolverhampton Borough Council (1978).
3. This approach is sometimes implied, for example, in the work of Herman Hertzberger. For an interesting account of this architect's ideas, see Hertzberger (1971).

Chapter 8.
1. For current British prices, see Spon (updated annually).
2. For information on *Visicalc*, see Software Arts (1980).

Suggestions for further reading

Introduction

The Responsive Environments approach to design starts from the idea that there are important relationships between social life and the arrangement of the built environment. *The social logic of space,* by Bill Hillier and Julienne Hanson, gives a comprehensive account of the spatial aspects of this relationship. Although it is not an easy read, this book is of great importance to all architects and urban designers.

It is also important for designers to understand why the qualities which support responsiveness are so often difficult to achieve in modern designs. In the end, this is largely due to the operations of the powerful economic interests which fund the property development industry: a thorough and well-written account of how this industry works is presented in David Cadman and Leslie Austin-Crowe's *Property development*. A vehement critique of the development process, well worth reading, is put forward in *The property machine,* by Peter Ambrose and Bob Colenutt.

There is a sad dearth of attempts to explain how this development system affects the pattern of physical form. Alison Ravetz's book *Remaking cities* is the widest available exploration of this topic, and should certainly be read; whilst Ian Bentley's short paper *Bureaucratic patronage and local urban form* focusses on the relationship between the investment aims of large financial institutions, the physical forms of city centre buildings and public places, and current architectural ideas. Though sketchy, the same author's *User choice and urban form* is interesting because it relates the impact of institutional funding directly to the qualities of permeability, variety and legibility.

Permeability

It is helpful to understand the historical roots of the interests and attitudes which currently combine to reduce permeability. *The fall of public man,* by Richard Sennett, gives an interesting historical analysis of this topic, with several references to buildings and urban places. A parallel argument, with less depth but easier to read, and related specifically to the physical environment, is contained in *The private future* by Martin Pawley.

The external factors which reduce permeability are reinforced by designers' own habits: particularly the pervasive tendency towards hierarchical spatial organisation. As an antidote to this attitude, read Christopher Alexander's essay *The city is not a tree,* together with Jane Jacobs's down-to-earth arguments for small blocks in Chapter nine of her *Death and life of great American cities.*

Working at a smaller scale, Stanford Anderson's forbiddingly titled *Studies towards an ecological model of the urban environment* puts forward a useful approach for considering permeability from public space into the block itself. Helpful advice on the detailed design of this transition is given in Stewart and Ricki McKenzie's article *Composing urban spaces for security, privacy and outlook.*

Variety

The social advantages gained from variety of use are explored by Richard Sennett in *The uses of disorder.* Not specifically about architecture or urban design, this book nevertheless uses many built-form examples, and makes stimulating reading.

There is a disappointing lack of literature about practical principles for achieving variety, though both Dmitri Procos's *Mixed land Use* and Eberhard Zeidler's *Multi-use architecture in the urban context* give lots of examples at various scales. These books are both worth looking through, if only because of the boost to morale which comes from seeing that variety *can* still be achieved.

Part two of Jacobs's *Death and life* supplements these examples with suggestions about how variety can be supported in physical terms, with particular reference to the role of old, low-rent building stock. In *Urban design as public policy* - a book by an architect who has worked hard to become street-wise in development economics - Jonathan Barnett suggests how relatively low-rent space can be created in *new* buildings by cross-subsidy. This is a seminal book, and is essential reading.

Barnett makes it clear that a good working knowledge of developers' methods of calculating financial feasibility is an important part of the designer's technical expertise for achieving variety. For a handy compendium of techniques and examples, which goes beyond what we ourselves have space for, see Architects' Journal's *Funding for construction*. A broader exploration of this topic is to be found in R.D.B. Booth's *Early perspectives in the valuation of property*. This is a clear, entertaining book. Written by a valuer, it scythes through the undergrowth of jargon which so often obscures the subject. It is also cheap enough to be on any designer's bookshelf.

Legibility

Our own rules of thumb for achieving legibility owe much to Kevin Lynch's pioneering book *The image of the city*. This should certainly be read: again, it is a cheap paperback which should be on every designer's shelves. Christian Norberg-Schulz, focussing on basically the same elements as Lynch, offers suggestions in *Existence, space and architecture* as to why these elements are important; and then illustrates them across a variety of scales. In addition, it is well worth reading Chapters 2, 3 and 4 of Barrie Greenbie's *Spaces,* for its interpretation - and illustration - of topics such as districts and paths.

For a wealth of historical examples of legible urban places - described measured and illustrated - it is hard to beat Camillo Sitte's *The art of building cities*. Long out of print, this is still available from libraries. Another ancient book long overdue for republication is Raymond Unwin's *Town planning in practice;* which has nothing to do with town planning as understood today. It is important because it observes and analyses historical urban places, deriving principles from them which it uses to achieve legibility in what, at the time, were new residential developments.

A more easily available book of historical examples is Edmund Bacon's *Design of Cities*. This is a lightweight general reader on the development of urban form, but has some useful explanations, together with copious illustrations of the use of publicly-relevant buildings to promote legibility. Beware of its emphasis on enormous urban interventions.

Useful reminders about the importance of small-scale factors in achieving legibility are given by Gordon Cullen in his *Townscape*. It is difficult to discern any real structure to this book, but its value lies in the meticulous observation and illustration of ways in which legibility can be achieved at a detailed level. In addition, *Townscape* argues strongly that designers should consider the *sequences* in which places will be experienced by their users.

Cullen often points to the role of signs, trees and so forth in achieving legibility. For the use of signs, try Robert Venturi's *Learning from Las Vegas:* the principles are usually more valuable than the examples here. For trees, read Henry Arnold's *Trees in urban design;* highly recommended for its advice on how to use trees to define, emphasise and subdivide urban spaces. There are many illustrated examples, mostly from the United States, plus technical backup on issues like maintenance, economics and soil characteristics.

One of the greatest enemies of legibility is the currently usual tendency for designers to treat all projects as though they were of equal public relevance. All too often, this leads to every unimportant office block being designed as though it were the city hall. An excellent cure for this disease is to read *The architecture of the city* by Aldo Rossi, which focusses on the crucial distinction between publicly relevant buildings and spaces, and the rest of the urban tissue. Beware, though: from the responsiveness point of view, Rossi's *ideas* are far more relevant than his own projects.

Robustness

Many books and articles have been written on the subject of 'user control'. Mostly, however, these consider the *patron* as the user. In practice, therefore, the design ideas they promote consist usually of gadgets to save space or devices to make building management easier. Few of these have any real potential for increasing robustness from the user's point of view. For a wide-ranging account of how technology feeds on this ambiguity, see *Mechanisation takes command* by Siegfried Giedion. *Housing flexibility*, by Andrew Rabeneck and others, gives a good survey of well-intentioned proposals; but most of them fall into the trap outlined above.

Moving out of doors, an exhaustive account of how outdoor places are used is given in William Whyte's *Social life of small urban spaces*. This small book is the culmination of many years spent observing and recording, particularly in New York. The work is well illustrated and easy to read, with many practical suggestions about the design and management of small public places.

Whyte's work gave rise to an organisation called Project for Public Spaces Incorporated. Operating in the USA, PPS undertakes detailed design and management projects to help public spaces support a diverse range of activities. As part of this effort, PPS has published many reports and films. The list is too long to give in full, but the address to contact for details is given in our bibliography.

Much helpful advice about achieving robustness in larger outdoor spaces, such as parks, is given in Albert Rutledge's inappropriately titled book *A visual approach to park design*. Much more than a merely visual approach is put forward here, and the book offers useful methods for observing and recording how public places are used.

Several useful ideas for the design of the interface between buildings and outdoor spaces, as well as for detailed design within outdoor spaces themselves, are suggested in Christopher Alexander's *Pattern language*. This puts forward 253 patterns for the design of the built environment, intended for use by lay people and professionals alike. Patterns 69, 88, 92, 93, 105, 106, 114, 119, 124, 125, 126, 140, 160, 164, 165 and 166 all offer interesting problem statements and suggested solutions to issues concerned with robustness.

In many situations, one of the key issues in achieving robust outdoor spaces involves working out ways of enabling pedestrians and vehicles to co-exist in comfort. In *The woonerf in city and traffic planning*, Dirk Grotenhuis gives very detailed information about the philosophy, design, management and legislation needed to achieve spaces of this kind in the Netherlands. This booklet will help readers to gauge the possibility of making similar advances in countries where planning and highways legislation is less supportive. In convincing the relevant authorities, it will be helpful to know about the detailed assessments of how these places have worked out in use, given by the Netherlands Ministry of Transport and Public Works in their report *From local traffic to pleasurable living*.

Visual appropriateness

Our whole aproach to visual appropriateness stems from the idea that definitions of 'good design' vary between different social groups, and that each of these different definitions is perfectly valid for the group which holds it. This is often quite difficult for professionally-trained designers to grasp. A useful first step for anyone who has this difficulty would be to read the clearly-written arguments put forward by Peter Berger and Thomas Luckmann in *The social construction of reality*.

The concept of reality as socially constructed is related

to the issue of aesthetics in Janet Wolff's *Aesthetics and the sociology of art*, which argues convincingly that matters such as aesthetics are strongly rooted in factors like class and cultural background. Wolff herself uses few specifically architectural examples, so it is also helpful to read Juan Bonta's *Architecture and its interpretation*, which gives a fascinating account of how radically different interpretations of the same buildings arise even *within* a specifically architectural subculture. Michael Thompson's *Rubbish theory* illuminates this situation by showing the financial payoffs which accrue to those trend-setters who can manage to *manipulate* the ways in which buildings are interpreted: buy it as a slum, and sell it as part of the national heritage.

Because people's cultural experience plays a crucial role in the interpretations they make, it is useful for designers to be aware of the processes by which class and educational differences delimit the culture to which people have access. This is explored in Pierre Bourdieu's article *The aristocracy of culture* and - only for the really keen - in Basil Bernstein's three-volume *Class, codes and control*. Neither Bernstein nor Bourdieu have much to say about the specifics of architecture or urban design, but parallels are not hard to draw.

The ways in which people's objectives affect their interpretations - both of things and of people - are clearly brought out in two books which have the advantage of being directly concerned with the built environment. The first is *Who needs housing?* by Jane and Roy Darke: in pages 120-140, they show how the different objectives of the various interest groups involved in housing provision affect their views of themselves and of the other groups involved. The second book, Matrix's *Making space,* is a trenchant interpretation of the largely man-made environment from a feminist point of view. Highly recommended.

The writings suggested above are fruitful sources of insights about the way interpretation works. But very little has been written about ways of applying these insights in design. *Does post-modernism communicate?* by Linda Groat and David Canter, is useful in a negative sense, in that it shows the fruitlessness of randomly mixing different architectural codes, as in much of the so-called post-modernism. In *Measuring the fit of new to old,* however, Groat begins to make positive recommendations, with some helpful advice about working with contextual cues.

A serious problem in achieving visual appropriateness arises from the unfortunate fact that designers have learned to denigrate the cultural mores of other groups, rather than taking them seriously. This makes it emotionally difficult for many designers to think clearly about the topic. In this situation, it may be psychologically easier to start by thinking it through in the context of a culture rather different from our own. Ian Bentley's *Design for a common cause,* for example, is an historical analysis of an approach very much like our own, as seen in the work of the Jugoslav architect Jože Plečnik. The argument is extended in the same author's *Jože Plečnik and the question of style.*

Richness

Though much has been written about the supposed importance of 'visual complexity', most of the arguments put forward are supported only by evidence from laboratory experiments in psychology. A rare attempt to check some people's reactions to more or less complex building surfaces is written up in Chapter five of *Meaning in the urban environment,* by Martin Krampen.

Many authors have commented at length about the twin dangers of over-complexity and over-simplicity, but here again there is rarely any useful guidance about what this might imply in practical terms. A wealth of analysis and examples, however, is contained in *The sense of order,* by Ernst Gombrich. This book is well worth reading for the way in which it makes clear the fact that because many visual events are experienced simultaneously, visual variety and visual order are interdependent.

Personalisation

Useful literature on personalisation is thin on the ground. There are many articles which exhort designers to encourage it, of which *Looking for the beach under the pavement* by Herman Hertzberger is perhaps the best. This gives a good introduction to the ideas of a designer who has constantly tried to support personalisation in his own work; but it contains no real practical advice.

Lars Lerup's *Building the unfinished* is again strong on exhortation, but does make some extremely interesting links between personalisation and visual appropriateness; suggesting the possibility of using visual cues positively to support personalisation. This is certainly worth reading, preferably *after* our own Chapters 5 and 7.

For the designer who wants to promote personalisation, one of the most important assets is a feel for what users are *likely* to want to do. An interesting account of what one set of users did (despite the designer's intentions) is given in Philippe Boudon's *Lived-in architecture:* Le Corbusier's Pessac half a century on. For discussion of the far less dramatic efforts which have been made to personalise the British inter-war suburbs, and the ways in which the original design supported these, see Ian Bentley's chapter *The owner makes his mark,* in *Dunroamin,* edited by Paul Oliver.

Finally, some of the most useful insights about the extent to which users are able to personalise can be gleaned from the pages of the many do-it-yourself manuals now available. One of the most comprehensive is that produced by Reader's Digest. Another, which puts particular emphasis on ways of personalising images, both internal and external, is *Decorating and beautifying your home,* in the Golden Hands series. Both are useful in developing a feel for how to make personalisation technically easier, to give the user's creativity a freer rein. That, in the end, is what responsiveness is all about.

Bibliography

Alexander, C. (1965) 'The city is not a tree' in *Architectural Forum*, 172, 1965.

Alexander, C. (1971) *Notes on the synthesis of form*, Cambridge (Mass): Harvard University Press.

Alexander, C. et. al. (1977) *A pattern language*, New York: Oxford University Press.

Ambrose, P. and Colenutt, R. (1975) *The property machine*, Harmondsworth: Penguin.

Anderson, S. (1981) 'Studies towards an ecological model of the Urban environment' in Anderson, S. (ed) *On Streets*, Cambridge (Mass): MIT Press.

Appleton, J. (1975) *The experience of landscape*, London (etc): Wiley.

Appleyard, D. (1969) 'Why buildings are known: a predictive tool for architects and planners' in Broadbent, G. et. al. (eds) (1980) *Meaning and behaviour in the built environment*, Chichester: Wiley.

Architects' Journal, The. (1984) Special number on *Funding for construction*, Vol. No. 22-29 August, 1984

Architectural Review (1977) *Enigma of the Rue Renard*, Vol. CLXI No. 693 pp. 270-294, May 1977.

Arnold, H.F. (1980) *Trees in urban design*, New York: Van Nostrand Reinhold.

Bacon, E.N. (1975) *Design of cities*, London: Thames and Hudson.

Baker, C.V. (1976) *Housing associations*, London: Estates Gazette.

Barnett, J. (1979) *Urban design as public policy*, New York: McGraw Hill.

Barrett, J. (1979) *The form and functions of the central area*, (Open University DT 201 Unit 12), Milton Keynes: Open University.

Bates, J. et. al. (1978) *Research report 20: A disaggregate model of household car ownership*, London: Department of Transport.

Bates, J. et. al. (1981) *The factors affecting household car ownership*, Farnborough: Gower.

Bathurst, P.E. and Butler, D.A. (1980) *Building cost control techniques and economics*, London: Heinemann.

Bentley, I. (1981) 'The owner makes his mark' in Oliver, P. (ed) *Dunroamin: the suburban semi and its enemies*, London: Barrie and Jenkins.

Bentley, I. (1983a) *Bureaucratic patronage and local Urban form*, (JCUD Research Note 15) Oxford: Joint Centre for Urban Design, Oxford Polytechnic.

Bentley, I. (1983b) 'Designing responsive places' in *Urban Design Quarterly*, London: Urban Design Group.

Bentley, I. (1983c) 'Design for a common cause' in Bentley, I and Gržan-Butina, D. (eds) (1983) *Jože Plečnik*, Oxford: Joint Centre for Urban Design, Oxford Polytechnic.

Bentley, I. (1984) *User choice and Urban form: the impact of commercial redevelopment*, (JCUD Research Note 18) Oxford: Joint Centre for Urban Design, Oxford Polytechnic.

Bentley, I. (1984) *Jože Plečnik: the question of style*, (JCUD Research Note 19) Oxford: Joint Centre for Urban Design, Oxford Polytechnic.

Berger, P. and Luckmann, T. (1966) *The social construction of reality*, Harmondsworth: Penguin.

Bernstein, B. (1971) *Class, codes and control*, (3 vols) London: Routledge and Kegan Paul.

Bonta, J.P. (1979) *Architecture and its interpretation*, London: Lund Humphries.

Booth, R.D.B. (1984) *Early perspectives in the valuation of property*, Oxford: Elsfield.

Boudon, P. (1972) *Lived-in architecture: Pessac revisited*, Cambridge (Mass): MIt Press.

Bourdieu, P. (1980) 'The aristocracy of culture' in *Media, Culture and Society*, 1980, 2, pp 225-254.

Bowyer, J. (1979) *Guide to domestic building surveys*, London: Architectural Press.

Broadbent, G. (1977) 'A plain man's guide to the theory of signs in architecture' in *Architectural Design*, 1977, 7-8, pp 474-482.

Brown, F. (1976) *Variety and Complexity in City Centre Renewal*, Oxford: Oxford Polytechnic, Department of Architecture major study.

Cadman, D. and Austin-Crowe, L. (1978) *Property development*, London: Spon.

Clarke, L. (1973) *Explorations into the nature of environmental codes: the relevance of Bernstein's 'Theory of Codes' to environmental Studies*, Working Paper No. 8, September 1973. Kingston: Architectural Psychology Research Unit, Kingston Polytechnic School of Architecture.

Cowan, P. (1963) 'Studies in the growth, change and ageing of buildings' in *Transactions of the Bartlett Society*, Vol VI.

Cullen, G. (1961) *Townscape*, London: Architectural Press.

Darke, J. and Darke, R. (1979) *Who needs housing?* London: Macmillan.

Department of the Environment and Department of Transport (1977) *Design Bulletin 32*, London: DoE and DoT.

Department of Transport (1966) *Roads in urban areas*, London: DoT.

Douglas, M. (1978) *Cultural bias*, Royal Anthropological Institute, Occasional Paper No. 35.

Douglas, M. (1982) *Essays in the sociology of perception*, London: Routledge and Kegan Paul.

Duffy, F. et. al. (1980) *Taking stock: a technical study on an approach to the undertaking of feasibility studies for the re-use of vacant industrial buildings*, London: URBED.

Filler, M. (1978) 'Extra sensory perceptions' in *Progressive Architecture*, Vol 59 No. 4, April 1978, pp. 82-85.

Garland, S. (1984) *The herb garden*, Leicester: Windward.

Geretsegger, H. and Peintner, M. (1979) *Otto Wagner 1841-1914*, London: Academy Editions.

Gibson, J. J. (1966) *The senses considered as perceptual systems*, Boston: Houghton Mifflin.

Giedion, S. (1948) *Mechanisation takes command*, New York: Oxford University Press.

Golden Hands Series (1972) *Decorating and beautifying your home*, London: Watts, Franklin.

Gombrich, E. (1978) *The sense of order*, Oxford: Phaidon.

Greenbie, B. B. (1981) *Spaces: dimensions of the human landscape*, New Haven: Yale University Press.

Grimsby Borough Council (1976) *House improvements and the street scene*, Grimsby: Borough Council.

Groat, L. (1983) 'Measuring the fit of new to old: a checklist on contextualism' in *Architecture*, Nov. 1983 pp. 58-61.

Groat, L. and Canter D. (1979) 'Does post-modernism communicate?' in *Progressive Architecture*, Dec. 1979 pp. 84-87.

Grotenhuis, D. H. (1978) *The woonerf in city and traffic planning*, Delft, Netherlands: Traffic Department, Public Works Services, Municipality of Delft.

Halprin, L. (1969) *The RSVP cycles: Creative processes in the human environment*, New York: Braziller.

Hanson, J and Hillier, B. (1982) 'Domestic space organisation: two contemporary space-codes compared' in *Architecture and Behaviour* 1982, 2, pp. 5-25.

Heery, G. T. (1975) *Time cost and architecture*, New York: McGraw Hill.

Hertzberger, H. 'Looking for the beach under the pavement' in *RIBA Journal* Vol 78, August 1971, pp. 328-333.

Hillier, B and Hanson, J. (1984) *The social logic of space*, Cambridge: Cambridge University Press.

Housing Corporation (1980) *Circular 11/80* London: HC.

Housing Corporation (frequently updated) *Housing Corporation Schemework Procedure Guide*, London: HC.

Jacobs, J. (1965) *The death and life of great American cities*, Harmondsworth: Penguin.

Krampen, M. (1979) *Meaning in the urban environment*, London: Pion.

Lerup, L. (1977) *Building the unfinished: architecture and human action*, Beverley Hills: Sage Publications.

Lynch, K. (1975) *The image of the city*, Cambridge (Mass): MIT Press.

Markus, T. (ed) (1979) *Building conversion and rehabilitation: designing for change in building use*, London: Butterworth.

Matrix (1984) *Making space: women in the man-made environment*, London: Pluto Press.

McKenzie, J.S. and McKenzie, R.L. (1978) 'Composing urban spaces for Security, privacy and outlook' in *Landscape Architecture*, September 1978 pp. 392-397.

McLaughlin, E. and McLaughlin, T. (1978) *Cost effective self-sufficiency, or the middle-class peasant*, London: David and Charles.

Miller, G. (1956) 'The magical number seven, plus or minus two: some limits on our capacity for processing information' in *The Psychological Review*, vol 63, March 1956.

Moore, G.T. (1983) 'Knowing about environmental knowing: the current state of theory and research on environmental cognition' in Pipkin, J.S. et. al. (eds) (1983) *Remaking the city: social perspectives on urban design*, New York: State University of New York Press.

Netherlands Ministry of Transport and Public Works (1983) *From local traffic to pleasurable living*, The Hague, Netherlands: MTPW Information Division.

Neufert, E. (ed) (1970) *Architects' data*, London: Lockwood.

Noble, J. (1983) *Local standards for the layout of residential roads: a review*, London: Housing Research Foundation.

Norberg-Schulz, C. (1971) *Existence space and architecture*, New York: Praeger.

Pawley, M. (1973) *The private future: causes and consequences of community collapse in the west*, London: Thames and Hudson.

Penwarden, A.D. and Wise, A.F.E. (1975) *Wind environment around buildings: a Building Research Establishment report*, London: HMSO.

Process Architecture (1978) *Lawrence Halprin*, Special issue, February 1978.

Procos, D. (1976) *Mixed land use: from revival to innovation*, Stroudsburg, Pa: Dowden, Hutchinson and Ross.

Project for Public Spaces Inc. (1982) *Effective pedestrian improvements in downtown business districts*, (Planning Advisory Service Report No. 368). New York: American Planning Association. (Note: PPS publications may be obtained from Project for Public Spaces Inc., 153 Waverly Place, New York, NY 10014, USA).

Rabeneck, A. et. al. (1974) 'Housing flexibility' in *Architectural Design*, vol 49 No.2 pp. 76-91, 1974.

Ravetz, A. (1980) *Remaking cities: contradictions of the recent urban environment*, London: Croom Helm.

Reader's Digest (1968) *Complete do-it-yourself manual*, London: Reader's Digest Association.

Rossi, A. (1982) *The architecture of the city*, Cambridge (Mass): MIT Press.

Royal Dutch Touring Club (1980) *Woonerf*, (2nd.ed.), The Hague: RDTC Traffic department.

Royal Institute of British Architects (1969) *Construction indexing manual*, London: RIBA Publications.

Rutledge, A. (1980) *A visual approach to park design*, New York: Garland STPM Press.

Sennett, R. (1970) *The uses of disorder: personal identity and city life*, Harmondsworth: Penguin.

Sennett, R. (1977) *The fall of public man*, Cambridge: Cambridge University Press.

Shils, E. (1981) *Tradition*, London: Faber and Faber.

Sitte, C. (1945) *The art of building cities*, New York: Reinhold.

Software Arts (1980) *VisiCalc user's guide*, Sunnyvale, Ca: Personal Software Inc.

Spon (updated annually) *Spon's architects' and builders' price book*, London: Spon.

Steinitz, C. (1968) 'Meaning and congruence of urban form and activity' in *Journal of the American Institute of Planners*, vol 34, pp. 233-247.

Thompson, M. (1979) *Rubbish theory: the creation and destruction of value*, Oxford: Oxford University Press.

Tutt, P and Adler, D. (eds) (1981) *New Metric Handbook*, London: Architectural Press.

Unwin, R. (1909) *Town planning in practice*, London: Fisher Unwin.

Venturi, R. et. al. (1977) *Learning from Las Vegas*, Cambridge (Mass): MIT Press.

Whyte, W. H. (1980) *The social life of small urban spaces*, Washington D.C: Conservation Foundation.

Williams, B and associates (1976) *Property development feasibility tables*, London: Building Economics Bureau.

Wolff, J. (1983) *Aesthetics and the sociology of art*, London: George Allen and Unwin.

Wolverhampton Borough Council (1978) *Streets ahead: a guide to improving the appearance of your house*, Wolverhampton: Borough Council.

Zeidler, E. H. (1983) *Multi-use architecture in the urban context*, Stuttgart: Karl Kramer Verlag.

Illustration credits

In addition to the authors' own illustrations, drawings or photographs were contributed by the following: Richard Anderson, pp 20-26; Gabriel Anzola-Wills, p 115; Graham Barnes, p 50 col 2 upper; Cristina Barrios, pp 82, 83 and 84 col 1; Anna Beauchamp, p 28 col 1 lower, Mita Bhaduri, p 79; Douglas Brown, p 33; Stefania Campbell, p 80 col 3; Simon Clark, p 131; Richard East, p 27 col 2 upper and lower, p 28 col 2, p 42 col 2 upper and lower; Richard Goddard, p 34 col 3, p 78 col 1; Brian Goodey, p 43 col 2 upper; Grimsby Borough Council, p 99 col 2; Teresa Heitor, p 50 col 1; Ideal Home, p 77 col 1 lower; Drew Mackie, p 6; MIT Press, p 43 col 2 lower, p 45 col 2 upper; Iradj Parvaneh, photographs, p 5; Ian Pope, help with photographic reductions; Ivor Samuels, p 50 col 2 lower; Thames and Hudson, p 44 col 1; Chris Trickey, pp 35-40. We are very grateful to them all.

Index

Books of related interest

Urban Design: Street and Square
Cliff Moughtin
A textbook and useful reference to the main
elements of urban design: public buildings, streets
and squares.
Spring 1992   224pp illus   220 × 220mm paper
ISBN 0 7506 0416 6

Landscape Detailing Second edition
Michael Littlewood
Specimen details for tracing or copying of the
commonest external works elements, from free-
standing walls to gates, steps and tree surrounds.
1986   224pp illus   297 × 210mm paper
ISBN 0 85139 860 X

Roadform and Townscape Second edition
Jim McCluskey
Aesthetic and urban values are combined with
effective engineering solutions to give valuable
guidance on the design of roads in relation to their
surroundings.
Spring 1992   344pp illus   297 × 210mm cloth
ISBN 0 7506 1245 2

The Concise Townscape
Gordon Cullen
The original classic on the art of giving visual
coherence to streets, layouts and spaces, and how to
perceive, enjoy and describe them.
1961   200pp illus   210 × 160mm paper
ISBN 0 85139 568 6   Not for sale in the USA

Tree Detailing
Michael Littlewood
A distilled compendium of practical information on
all aspects of tree planting and maintenance.
1988   224pp illus   297 × 210mm paper
ISBN 0 408 50002 6

Manual of Graphic Techniques Volumes 1–4
Tom Porter and Sue Goodman
Practical handbooks, each one complementing the
others.
Vol 1: Materials and step-by-step methods
Vol 2: The image-making process
Vol 3: Wider design techniques
Vol 4: Development and presentation of plans
Each volume 128pp illus   210 × 297mm paper
Vol 1 (1980) ISBN 0 408 50012 3
Vol 2 (1982) ISBN 0 408 50007 7
Vol 3 (1983) ISBN 0 408 50008 5
Vol 4 (1985) ISBN 0 408 50009 3

Crime Free Housing
Barry Poyner and Barry Webb
Design criteria are applied to reach a clearly
established set of design recommendations for
layout, access patterns and spatial relationships.
1991   144pp illus   246 × 189mm paper
ISBN 0 7506 1273 8

London Docklands: Urban Design in an Age of De-regulation
Brian Edwards
What type of urbanism results from the largest urban
waterside development project in Europe, when it is
also the least controlled through conventional
planning methods?
Autumn 1992   160pp illus   297 × 210mm cloth
ISBN 0 7506 1298 3

The City of Tomorrow
Le Corbusier
Le Corbusier's early vision, which was to have
tremendous influence on the 1920s and 30s.
1971   320pp illus   203 × 137mm paper
ISBN 0 85139 124 9   Not for sale in the USA

Complexity and Contradiction in Architecture
Robert Venturi
A scholarly challenge to the Modern Movement,
attacking 'the frustration of abstractly conceived
architectural order'.
1977   136pp illus   279 × 203mm paper
ISBN 0 85139 111 7   Not for sale in the USA